T0043647

Science and Fiction

Science and Fiction—A Springer Series

This collection of entertaining and thought-provoking books will appeal equally to science buffs, scientists and science-fiction fans. It was born out of the recognition that scientific discovery and the creation of plausible fictional scenarios are often two sides of the same coin. Each relies on an understanding of the way the world works, coupled with the imaginative ability to invent new or alternative explanations—and even other worlds. Authored by practicing scientists as well as writers of hard science fiction, these books explore and exploit the borderlands between accepted science and its fictional counterpart. Uncovering mutual influences, promoting fruitful interaction, narrating and analyzing fictional scenarios, together they serve as a reaction vessel for inspired new ideas in science, technology, and beyond.

Whether fiction, fact, or forever undecidable: the Springer Series "Science and Fiction" intends to go where no one has gone before!

Its largely non-technical books take several different approaches. Journey with their authors as they

- Indulge in science speculation—describing intriguing, plausible yet unproven ideas;
- Exploit science fiction for educational purposes and as a means of promoting critical thinking;
- Explore the interplay of science and science fiction – throughout the history of the genre and looking ahead;
- Delve into related topics including, but not limited to: science as a creative process, the limits of science, interplay of literature and knowledge;
- Tell fictional short stories built around well-defined scientific ideas, with a supplement summarizing the science underlying the plot.

Readers can look forward to a broad range of topics, as intriguing as they are important. Here just a few by way of illustration:

- Time travel, superluminal travel, wormholes, teleportation
- Extraterrestrial intelligence and alien civilizations
- Artificial intelligence, planetary brains, the universe as a computer, simulated worlds
- Non-anthropocentric viewpoints
- Synthetic biology, genetic engineering, developing nanotechnologies
- Eco/infrastructure/meteorite-impact disaster scenarios
- Future scenarios, transhumanism, posthumanism, intelligence explosion
- Virtual worlds, cyberspace dramas
- Consciousness and mind manipulation

More information about this series at http://www.springer.com/series/11657

Massimo Villata

The Dark Arrow of Time

A Scientific Novel

Massimo Villata
Osservatorio Astrofisico di Torino
INAF
Pino Torinese (TO)
Italy

ISSN 2197-1188 ISSN 2197-1196 (electronic)
Science and Fiction
ISBN 978-3-319-67485-8 ISBN 978-3-319-67486-5 (eBook)
https://doi.org/10.1007/978-3-319-67486-5

Library of Congress Control Number: 2017953782

The author published the Italian edition in 2012 with a selfpublishing, named ilmiolibro.
EAN code: 9788891010728

Printed on acid-free paper

This Springer imprint is published by Springer Nature
The registered company is Springer International Publishing AG
The registered company address is: Gewerbestrasse 11, 6330 Cham, Switzerland

Contents

Chapter 1
Very Fidgety, the Fat Lady Next to Him

Very fidgety, the fat lady next to him. Gripping the armrests tightly, every once in a while she would jerk her head a bit to one side or the other, for no clear reason. It was obviously her first trip, and you could tell from the blank look on her face, blank but strained, uncertain.

It was the first time for him too. It's not that he wasn't afraid, he was just trying to hide it, even from himself. And he was managing pretty well.

To some extent, actually, he was modeling himself on the serious man two seats down, who—at least apparently—was impassibly, calmly absorbed in his own thoughts, as if he did this every day, all just part of the job.

On the other side, a bit farther down, a black man was murmuring some kind of litany, like a subdued propitiatory chant.

They were all on a single, slightly curving row of seats, one beside another. He was almost in the center. Fifteen people, one seat was empty.

It was almost nine o'clock, almost time for liftoff before the transmission. The cabin with the sixteen seats at the center of the big room full of machinery was closed, and the enormous hatch in the ceiling began to open. It would just be a few minutes more, long enough to finish docking.

It wasn't a real trip. As he had learned in college, it wasn't a question of moving people or things from one place to another, at a certain speed and taking a certain amount of time. That wouldn't have been remotely feasible. Where he was going, around six parsecs[1] away, a trip would have taken several years at the speeds that could be achieved, even allowing for the relativistic contraction of distances.

By now, real trips were only used for short distances, inside the solar system or its immediate neighborhood. But even they were about to be made obsolete by the recent advances in the transmission technique, cheaper in terms of energy though still riskier, relatively speaking. A probability of around three per thousand of being lost for good in cosmic space, he'd read somewhere.

So it wasn't a trip, then, really just a hop, skip and a jump. Skipping a lot of trouble, too.

---

[1] 1 parsec $\approx$ 3.26 light-years.

M. Villata, *The Dark Arrow of Time*, Science and Fiction,
https://doi.org/10.1007/978-3-319-67486-5_1

Our common existence is played out in a space-time continuum where moving from one point to another necessarily takes a time which is positive and greater than zero. How much greater depends on the speed. In theory, high speeds close to that of light make it possible to reduce the amount of time needed at will. Any given distance can be covered in a time less than that taken by light, even though the speed of light, as we know, cannot be reached or exceeded in our 'world'. What changes is the passage of time, and the measure of the distances. If a spaceship travels at a velocity close to the speed of light, say at $v = 0.995\ c$,[2] over a distance (measured from Earth) of ten light-years, from our point of view it will take slightly over ten years to complete its trip. For the astronaut, on the other hand, the distances are significantly shorter, by around a factor of ten,[3] and so the travel time will also be shorter. Consequently, the normal, real interstellar voyages last for small fractions of the astronauts' lives, while decades go by back on Earth.

They are almost always one-way trips, by whole families or social groups setting out to colonize new planets. News about the trips arrives sporadically, and how they turned out will only be known many years later. The whole business, in any case, involves monumental wastes of time and energy.

Transmission was a real revolution. It is not a trip in our world. Somebody had dubbed it "riding the light".

Fifty-five seconds. Fifty-four. Fifty-three. The fat lady vacillates between apparent calm and attacks of the jitters. The buzzing resonates more loudly, its frequency increasing. Now it is almost a hiss. Or that's how it seems, at least. The sleeping gas pumped through the mask begins to take effect, and it is hard to tell what is really going on. Through the portholes, he glimpses the attendants checking the seatbelts, the masks and the monitors one last time. The seats tip back. The fat lady goes limp. Helias turns his head to look at the serious man, still imperturbable, though his face behind the mask seems to lengthen and change color.

The long face is in front of him now, mask off and eyes half closed. Then it starts to revolve around the tip of the nose, slowly at first, then whirling faster and faster. So fast it is just concentric circles, light and dark. Slowing again, it comes to a halt, upside down. Now it is wearing a pair of dark glasses, oval lenses fitting close to the face. "We've arrived" it whispers to him, the voice calm and reassuring. The fat lady is still sleeping and looks like she'll probably be at it for a while. Better let her wake up in her own good time. The black man is already on his feet, ramming on the eyeglasses handed to him and striding off toward the open door of the cabin, swinging his suitcase purposefully. Four people on the right are still sleeping. Some of the seats are already empty. A blond girl starts to stir, while Helias's seat returns to the upright position and he puts his feet down. The seatbelts unlatch automatically. A bit hesitantly, he gets up, finds his luggage, activates the flotation unit and

[2] $c \approx 299{,}800$ km/s is the speed of light.

[3] The distances contract by a factor $\gamma = (1 - \beta^2)^{-1/2}$, where $\beta = v/c$. In the case considered here, $\beta = 0.995$, so we will have $\gamma \approx 10.01$.

starts for the exit, already wearing his glasses. The serious man seems to be waiting for him over there, leaning on a machine in the open-ceilinged concourse.

Helias looked up at the sky. It seemed dark blue, but maybe that was because of the concentrator lenses in his glasses. There wasn't much visible light on that planet. But its sun emitted plenty of infrared. The glasses increased the frequency of the infrared photons, moving them into the optical band where the human eye could 'see' them. Though this resulted in having enough photons to be able to see, the colors were obviously muddled, a sort of dusky, artificial Daltonism in the gloaming.

He walked past the serious man, pretending not to see him but sneaking a look through the dark glasses. The quiet voice made itself heard again.

"Dr. Helias Kadler?"

That voice, calming and disquieting at the same time.

"Yes? Yes, I'm Kadler."

"I know you are going to the Kusmiri Center. If you wish, I can take you there."

It was all very strange. And so who was this man? What did he know about him and his 'call'? He hadn't said, "I'm here to take you to the Center." But "If you wish, I will take you there". So he wasn't there to welcome him. How could he have been, in any case, since he had arrived from Earth with him? His trip, for him, had gone by in a flash, an instant. But on Earth, twenty years had passed. If he had immediately gone back to Earth, or rather, if he had been 'retransmitted', he would have been home again an instant later, but forty years after his departure, and with no guarantee that things would still be the way they were before. No wonder that transmissions beyond a certain distance were, at least these days, always one-way trips.

And so how could that imperturbable man have arrived together with him from Earth, and at the same time offer himself as a guide to the place? But maybe it was just his impression, since the man been so nonchalant about the transmission, and seemed so much at ease on that planet. Maybe he was a newcomer like him, but simply was well informed about the local topography or means of transportation. But even so, how did he know his name, and why did he offer to accompany him?

"How do you know my name?"

"I am called Mattheus Bodieur."

He hadn't answered the question, but at least he had introduced himself.

"Are you from here, sir?"

Before responding, he glanced around as if he wanted to make sure he wasn't overheard.

"Yes."

He was lying. If not, he would have had to have left at least forty years earlier, apparently for the sole purpose of 'accompanying' a person—him—who at the time was twenty years younger. Just a child, in other words.

Helias's first impulse was to say goodbye to the lying stranger and go his own way, following the directions given to him before departure. But his curiosity was piqued, and with his temperament, he wasn't about to leave without getting to the bottom of this. He stood looking fixedly at the serious man's dark glasses, noting that his skin seemed brown.

"If you need to think about it, may I at least have the pleasure of offering you a real Alkenian ice cream at the station emporium? It's just what you need as a pick-me-up after the 'jolt'."

A 'jolt', he called it.

"Since you are so kind…. Thank you."

The Alkenian ice cream was superb, he had never eaten one like it. By comparison, the ones on Earth were pale imitations. It really had been a pick-me-up, and now he felt more relaxed and in tune with that unknown planet.

"Are you really from here?"

"Yes."

"And when did you leave for Earth?"

No answer.

"You would have had to have left at least forty Earth years ago. Incidentally, how long does an Alkenian year last?"

He knew perfectly well. It was just to keep the ball rolling.

"Around 407 Alkenian days, that's 378 Earth days. There's not much difference with Earth, and so it doesn't take long to get used to it."

"Why do you want to accompany me to the Center?"

Silence.

"How do you know my name?"

"It was on the passenger manifest."

"Did you offered to accompany all of the passengers?"

"No."

"And so you left here at least forty Earth years ago. You arrived on Earth, you found that I was coming here and you decided to accompany me? Doesn't that strike you as absurd?"

"More or less."

"That's not a very precise answer, wouldn't you say?"

He was silent for a moment, and then murmured, almost to himself, "There's the trick.".

"What trick?"

The question was ignored. He finished his ice cream meticulously and went to pay. Then he motioned to Helias to go out with him and in the doorway, putting his glasses back on, said "You will know in due time.".

Any electromagnetic signal, such as light pulse or, for the sake of simplicity, a single photon, travels in empty space—a vacuum—at a constant speed c of around 300,000 km per second. This speed is independent of the motion of the light source and of the motion of whoever or whatever receives the photon. All observers who measure the speed of light, regardless of their state of motion, will always measure c.

Though it seems to contradict what we think we know about "everyday physics", this is the way things work. If a projectile—a bullet, say, is shot from a moving vehicle and in the direction of the vehicle's motion, its speed will be greater than if it were shot from a stationary position. Likewise, if the target is moving toward the bullet, the impact will be received at a higher speed than if it were moving away from the bullet's motion. And if the target is moving away at a higher speed than the bullet, the bullet will never catch up with it. But photons are not like bullets: with photons, you can run but you can't hide. Even if you try to escape, at any speed, the photon will always reach you with velocity c.

If we emit a photon from Earth, one second later it will be at distance of around 300,000 km. If at the moment of emission a spaceship passes at any given speed, half of the speed of light, for instance, the photon will also have traveled 300,000 km in one second from the astronaut's standpoint, and will thus be 300,000 km ahead of the spaceship, if the latter was going in the same direction as the photon. This would mean that the photon is simultaneously located 300,000 km and approximately 450,000 km from Earth, given that in the meantime the spaceship has traveled around 150,000 km. Clearly something doesn't add up here. In reality, the answer is quite simple, though it might seem strange. What changes is the passage of time, and the measure of the distances. A second measured on board the spaceship is 'different' from one measured on Earth. And the distance of 300,000 km measured from the Earth is no longer the same if it is measured from the spaceship. Yardsticks and clocks, space and time, change to adapt to the true universal constant, the speed of light.

We usually don't notice, because we deal with small velocities for which the relativistic contraction of distances and dilation of time are too small to be measured.

And so time is not an invariant. It does not flow smoothly and undisturbed, uniform and identical for everyone. It depends on the state of motion of the observer who measures it. We can make a space voyage at enormous speed, and come home to find that our son is twenty years older than us. And dimensions in space are not invariant either: objects and distances contract in the direction of motion, and their measurements depend on the relative velocity at which we measure them.

An elementary time interval is designated as $\mathrm{d}t$, while $\mathrm{d}r$ is a piece of space, an element of distance. Neither are invariant under Lorentz transformations. Lorentz transformations make it possible to calculate the space-time coordinates of an event in a given inertial frame, starting from the coordinates for the same event in another inertial frame moving at a constant velocity v relative to the first. For small relative velocities, $\mathrm{d}r$ and $\mathrm{d}t$ are almost invariant, as they appear in our everyday experience. For relative velocities approaching the speed of light, their variation from one frame to the other can become arbitrarily large. A length will contract by a factor $\gamma = (1 - \beta^2)^{-1/2}$ (where $\beta = v/c$), and time will slow by the same factor.

The simplest relativistic invariant, i.e., invariant under Lorentz transformations, is the square of the space-time four-vector, $\mathrm{d}s^2 = \mathrm{d}r^2 - c^2\mathrm{d}t^2$. In other words, whereas for Galilean physics (which still applies for small velocities), the invariants were $\mathrm{d}r^2$ and $\mathrm{d}t^2$, in relativistic physics, the invariant is a linear combination of the two, where the coefficient c (speed of light) is a universal constant. It should be noted that for a photon we have $\mathrm{d}s^2 = 0$, since $\mathrm{d}r$, or in other words the distance traveled, is equal to the product of speed, c, by the time $\mathrm{d}t$ taken to cover the distance. In this case, the space-time interval $\mathrm{d}s$ is said to be light-like. Likewise, it is called a time-like interval if $\mathrm{d}s^2$ is less than zero, i.e., when the spatial distance $\mathrm{d}r$ is less than that which light can travel in time $\mathrm{d}t$, meaning that the distance can be covered in time $\mathrm{d}t$ with a speed less than c, and consequently the two events at the end points of $\mathrm{d}s$ can be linked by a cause-effect relationship. By contrast, because no matter and nothing that transports energy or information can have a speed greater than c, the two events at the end points of a space-like interval, i.e., with $\mathrm{d}s^2$ greater than zero, cannot be causally linked, and in other words take place independently of each other.

We have just said that no matter, i.e., nothing with mass greater than zero, can exceed the speed of light. In reality, it cannot even reach the speed of light. Only an electromagnetic wave, or its quantum correspondent, the photon, which has null mass,[4] can, or rather, 'must' travel at speed c.

And so we have this apparently random coincidence: particles with $m = 0$ also have $\mathrm{d}s = 0$. A close relative of $\mathrm{d}s$ is the so-called 'proper time', $\mathrm{d}\tau$, where $\mathrm{d}\tau^2 = -\mathrm{d}s^2/c^2 = \mathrm{d}t^2 - \mathrm{d}r^2/c^2$. It is referred to as proper time because it is the temporal distance between two events that take place in the same point in space, i.e., with $\mathrm{d}r = 0$; in this case, in fact, $\mathrm{d}\tau = \mathrm{d}t$. In other words, it is the time measured with a clock at rest in the considered reference frame, hence 'proper time'.

A photon thus has $m = 0$ and $\mathrm{d}\tau = 0$ (since $\mathrm{d}s = 0$). Normally, we say that a photon does not have a reference frame, as the second postulate of the special theory of relativity (Einstein 1905) states that light travels at speed c in any reference frame. Consequently, there can be no such thing as a reference frame in which the photon is at rest.

The photon, whatever it is, is certainly something very strange and at the same time fundamental, that our description of physical phenomena can barely hint at. It is a sort of singularity of the physical world, a boundary that cannot be crossed, or at least apparently. It has null mass, as if it did not belong to our world, and null proper time, as if it were unaffected by the flow of time. For the photon, our space does not exist, because the closer speeds come to c, the more space contracts and tends to zero. So in a way it is as if it occupied the entire space in a single timeless event. No sooner has it left than it has already arrived, even if from our point of view it has traveled for billions of years and crossed half of the universe.

Consequently, having no mass is the essential, necessary and sufficient condition for traveling at the speed of light and not being affected by the passage of time.

And so how could Helias Kadler and the other travelers, with the whole spaceship and its far from negligible mass, "ride the light"?

[4]Or any other particles with null mass.

Chapter 2
They Were Flying Over Gentle Crimson Hills

They were flying over gentle crimson hills dotted with orange shrubs casting long violet shadows in the brooding light that announced the coming sunset. The air was perfectly clear and a brisk wind from the southwest shook the treetops in the woods at the foot of the hills, the tiny leaves shimmering in a broad swath of vibrating color, now yellow, now purple, now maroon, as the wind turned their different surfaces toward the watching eye. The undersides of the leaves were at their brightest yellow when the wind brandished the branches against the sun.

They were traveling north, almost along the planet's shadow line. At dusk, they had to increase altitude to clear a mountain range. And then the sun was back, tingeing with pink the immaculate snowfields and the plumes and pennants of clouds floating from the highest peaks. And where the sun failed to reach, the snow was blue and violet in the shadows, much like on Earth.

For a long time, Helias stood rapt before these breath-taking landscapes, forgetting all the problems and questions crowding his mind.

Only when darkness finally covered the land below and the first, brightest stars began to appear in the purple and green sky, did he turn from the porthole. Removing his glasses, he looked around the spaceship, his eyes growing accustomed to the dim white light. It was an eight-seater. Two rows of two seats per side, with an aisle down the middle. At the back, the baggage hold, at the front the cockpit with the two pilots. The door to the cockpit was ajar, and from his seat in the first row Helias could see the profile of the serious man, Mattheus, busy in the captain's seat. He was talking into the tiny mic of his 'cell', a miniscule earphone clinging to his earlobe and an invisible rod that supported the mic just to the right of his mouth. He was almost whispering, in a strange dialect. A few more words, interspersed with pauses as he listened, and then what seemed to be a goodbye of some kind and the conversation came to an end.

On the other side of the cabin, in the second row, a blond girl was drowsing. Hadn't he seen her before, somewhere?

Helias fell back into his own thoughts.

M. Villata, *The Dark Arrow of Time*, Science and Fiction,
https://doi.org/10.1007/978-3-319-67486-5_2

He had been 'called', after more than a year on the waiting list. He'd had himself put on the list right after he got his PhD. Shortly before, his girlfriend had left him. He had told her about his plans to leave Earth, and that he wanted to take her with him. She seemed reluctant to go: too attached to her family and her own habits. For her sake, he might even have decided not to leave. But as it turned out, she took things into her own hands, and made the decision for him, telling him it was over—just before his final exam. She didn't feel she should have to wait until he finally made up his mind, and in any case, she certainly didn't want to have to blame herself if he gave up his dreams of the future for her.

For his part, he had no family left. Or almost none. There was a sister, somewhere in the Austrian Alps. They had never had much to say to each other, and from the time she married and moved to Austria he had never found the time—or the inclination—to go visit. They heard from each other from time to time, birthdays or holidays and things like that, but nothing more.

For the few times he permitted himself a vacation, he preferred the Swiss or Italian Alps. The high peaks and the eternal snows, with his girlfriend. Or Corsica, which he'd always loved, ever since he went, as a boy, with his parents.

His parents were gone, officially declared missing. They had been two prominent scientists, who worked at the transmitting station orbiting Mars.

He hadn't seen much of them, since he started college and was living on his own. They would come and go, staying for a while, busy with their studies, and then leave again. Once they had taken him with them. It was during the school break, and there was an extra seat available on the spaceship. He was sixteen, and he remembered it as the best period of his life.

Since then, he had had fewer and fewer chances to see them, since they spent increasingly long periods on Mars. Until two years before he finished his PhD. They didn't come back from their last trip. He never heard from them again.

Helias's eyes filled with tears, and his throat tightened.

But this was no time to get sentimental. And he was used to pulling the plug on his feelings, and pushing everything back down, deep underneath.

He allowed himself one small liberating sob, and slowly took control of his thoughts again.

He turned for a moment, glancing back at the seats behind to see if the girl had heard anything and was watching him. Nothing, she seemed completely absorbed, her eyes half closed.

For good measure, he pretended to cough, just to belie any suspicions.

He turned again and gazed at girl's face, as her lips seemed to move almost imperceptibly. It was an oblong face, though not too much so. The eyes seemed narrow, maybe because they were half closed, almost like an Oriental. Despite the fact that they were blue, and despite the fine blond hair that fell to her shoulders. She had a goodish figure, on the tall and slender side, though the loose coveralls made it hard to tell. Not his type, he told himself, though he couldn't deny a certain attraction.

The girl turned toward him, and he looked away immediately.

And now he found himself in this strange, unexpected situation.

He had been directed to the Kusmiri Center, where, among other things, they did research into alien molecular biogenetics. And where he would have been able to catch up with the field and start something new.

The last concrete news he had about the Kusmiri Center and, in general, the planet Alkenia, obviously dated back some forty years, though he had heard it slightly before his departure. He hoped nothing fundamental had changed, even if he had to expect that there had been a great deal of progress.

The information exchanged between the two planets was always twenty years out of date, and was more of historical interest than of any value as information.

Above all, he knew he would have to assimilate forty years' worth of new biogenetic research, and he had no idea how much progress the discipline had made.

Back on Earth, there were simulations that described all aspects of social life and the organization of research on Alkenia in a certain amount of detail. They were based on successive transfers of personnel, information and plans that Earth sent at fairly close intervals, every few weeks or months. Naturally, the simulations were only relatively useful, since they were little more than forecasts starting from the planet's actual situation as it was twenty years earlier. And they could be confirmed or rejected only twenty years later.

Up to the time of Helias's departure there hadn't been any major surprises.

But in the meantime, forty years had gone by. And more than anything else, it was practically impossible to make predictions about the advances in research that used 'raw material' that didn't exist on Earth, except in the form of samples taken decades earlier.

According to the program he'd been given on Earth—every detail of which he had committed to memory—on leaving the station he was to take the shuttle to the terminal in the nearby city of Symiria, where he was supposed to take the first flight for the planet's capital, which wasn't far away. Once in the capital, he would have spent the night at the Hotel Starcross, taking the shuttle for the Kusmiri Center the next morning.

None of this happened. That serious-looking man had immediately blown a hole right through the entire program. He had barely introduced himself, without even saying who he was or what he wanted from Helias. To all appearances, he just seemed to be a nice man who wanted to give him a lift. But his caginess about offering any kind of explanation, and the strange circumstances surrounding their meeting had something mysterious about it that piqued his curiosity. And there was no denying that the man exerted a certain fascination over him, a kind of charisma.

Upon leaving the emporium, Mattheus had walked off toward the parking lot, without another word. And this attitude had irked him again. The man couldn't treat him like this, like a child that didn't deserve an explanation. He had been about to leave without even saying goodbye, but then he ended up trotting obediently along after him, because he 'felt' he had to get to the bottom of this.

The blond girl and the copilot were waiting on the spaceship. Embarking, his impulse was to sit next to the girl, if only in the hope that she might be more talkative than Mattheus. But she hadn't even turned to see who was coming in and—no surprise—she also seemed caught up in her own thoughts. So he had given up and, a bit huffily, went and sat as far away from her as he could.

While he was waiting for the flight to arrive, Helias reached into the inner pocket of his jacket and pulled out his little portable, which contained everything that could be digitized, from his childhood memories to all of his studies and research. He glanced at the index of his scientific publications, thinking that by now they were all old, obsolete analyses. With a bit of nostalgia, he looked at pictures from when he was a boy, the favorite photo of his parents, his ex-girlfriend. Pensively, he thought about how everything fades away and disappears, now more than ever before. He had no chance of going back to Earth, but even if he could and had wanted to, what would he have found down there? A seventy year old sister, aunts, uncles and all his relatives long dead, his former girlfriend with white hair, maybe surrounded by half a dozen grandkids and the children from her second marriage. With her first husband's portrait hanging on the wall, the first husband who, irony of ironies, was lost in space. And almost no memories remaining of their jaunts in the Alps, still so vivid in his mind.

He was beginning a new life, in every way. With no ties to the past, nothing more than a jumble of memories, old ones by now, and a few pieces of research that had since become meaningless.

In the midst of these thoughts, a light from outside caught his attention. A star, brighter than the others, was shining out among the low clouds on the horizon. It was Nasymil, the nearest star, now reflected imperiously on the clouds below, as the spaceship climbed to hurdle the last mountains. It was still mirrored in an immense glacier, brighter than the full moon. Then the lights went out in the spaceship as it began its descent toward the Center.

This new light shining on his thoughts of a new and unknown life struck him as a good omen, and he cheerfully prepared for the landing.

The Center appeared suddenly below him, no longer hidden by the looming mountain. And he had plenty of time to admire the architecture as the ship looped around it before landing. In the cold light of Nasymil, which was reflected now on the roofs of the towers, contrasting with the orange and yellow of the artificial lighting outside the building. It was enormous, a stupendous castle overlooking the choppy waters of a lake.

Despite his good mood, even better after seeing that fairytale landscape, and despite his eagerness to disembark and throw himself into his new life, he decided to keep his feelings to himself, given the lack of interest his traveling companions had shown in him. He was all ready to leave, but stayed glued to the porthole—not that there was much to see anymore—waiting for the girl to go out first. He got down his luggage and—sulkily, to all outward appearances—followed his two fellow travelers, while the copilot remained on board. Nothing, not even a word. What kind of a way to behave was that? Mattheus had barely glanced over his shoulder to check whether he was following. He went first through an ordinary sliding door and walked toward the reception area. Oh! A miracle! Once through the door, the girl seemed to slow down to let him catch up, and she was even turning toward him. With a smile!

"My name's Kathia."

"I usually don't introduce myself, since people already seem to know who I am anyway."

And he gestured with his chin toward Mattheus who had stopped at the reception desk.

They stopped too, a few paces behind.

She looked him in the eye and smiled. He felt himself thawing.

"Pleased to meet you, Mr. 'Usually-I-don't-introduce-myself'."

She held out her hand. He followed suit, and was about to shake hands when he realized that it clearly was not the custom here. In fact, she didn't shake hands either, but remained with her hand outstretched and open. He brought his palm close to hers and felt something like a halo of warmth. Foolishly embarrassed, he snatched his hand back. She smiled again, almost maternally. She continued to look at him with an amused air, like someone bending over a new-born puppy, fuzzy but still clumsy. To hide his embarrassment, Helias asked, "Are you here for bio-genetics too?".

"No, archivist."

Nothing else? Why were people in this place so close-mouthed? Not that he was particularly loquacious himself, but these two had him beat by a country mile.

Mattheus came back to them with a couple of passes.

"These are temporary, the room number is at the top. They're on the other side of the building. The registration office is closed now, obviously. You're expected for the formal registration tomorrow morning, in the office alongside. The rooms should be in order. If you need anything, call Six. Rest well."

Wow! What a spiel! And almost without stopping for breath.

Mattheus smiled. He looked at Helias with a penetrating gaze. Then he looked at Kathia.

They said goodbye to him, went out of the building and made their way toward the opposite side, with a little detour to walk along the lake.

"What a strange fellow...." murmured Helias. But he got no reaction.

"Had you already met him?"

"When?"

"Before today, I mean."

A short silence.

"Yes."

"Do you come from Earth too?"

"Sure."

"What I mean is, have you just arrived, like me?"

"We were on the same craft."

Four whole words in a row! He noticed she had an odd accent.

And so that's where he'd seen her.

Why was it so difficult to engage her in a conversation? Maybe she had problems with the language?

"Where do you come from?"

"Sweden."

With that accent? Nonsense. Why was she lying, her too?

He didn't know what else to say, he was too discouraged. He threw out a stupid question, for lack of anything better.

"How did the 'trip' seem to you, the transmission, I mean? I was pretty excited...."

"I acknowledge that."

I acknowledge that?! Craft?! What kind of a way to talk was that? His discouragement was turning to exasperation. A Swede with a Spanish, or maybe Portuguese, accent. To hell with it all! Fortunately they'd arrived at the entrance with the numbers of their rooms.

They went up four flights of wooden stairs. And through a number of corridors. From 331 to 335. It was hers.

"You've arrived."

"Yes, I see."

He made one last try, since a million questions were gnawing at him.

"You don't have a Swedish accent. Tell me the truth, where do you come from?"

She smiled again, a disarming smile.

Then she half closed her eyes, put on a sober, level-headed expression and with a perfect European accent pronounced, "You will know in due time.".

A shiver ran down Helias's spine and he felt that his hair was standing on end. It was a perfect imitation. The same words Mattheus had used.

The girl started soberly toward her room, but after a couple of steps, without stopping, she turned, winked at him, and swaying her hips disappeared behind the corner.

With the most idiotic possible look on his face, Helias stood as if nailed to the floor.

These people were teasing him.

He couldn't stand it.

He really couldn't stand it.

Chapter 3
Helias Slept Fitfully

Helias slept fitfully. His dreams had never been so confused, or so crowded with people. It was as if he were trying desperately to mix people from his past with those from the present, his only possible present.

Half asleep, he again saw Mattheus who, leaving, shook his hand. Only his, not Kathia's. Why had he shaken his hand, while the girl had held hers open? Then he sank back into slumber, and Mattheus had become his father, older now, who was looking at his exam results and smiling at him, as he squeezed his arm affectionately. But then he turned back into Mattheus, who with a penetrating glance told him, "You have to discover it on your own, there's no room for idiots on this planet." And then the dreams became more and more confused, mixing people from other times, as if they wanted to survive, wanted not to be forgotten. Once in a while, Kathia's face appeared clearly, smiling at him with her Oriental eyes, now blonde, now brunette, but always smiling and consoling, with a maternal air.

The counselor paced back and forth in the big room. He stopped at intervals, as if trying to stop a thought, stroking his bearded chin.

"In conclusion, what do you think of the young man?"

Mattheus stalled, walking over to stir the fire in the enormous hearth.

"He's bright, no doubt about that."

"Don't evade the issue, you know perfectly well what I mean. Is he how they described him?"

"I think so. But I'd like to check some more. Kathia listened to him more than I did. I have to talk to her. Tomorrow."

Mattheus took his leave and the counselor returned to pacing back and forth.

Helias woke late, without understanding whether he was really awake or still dreaming. Bit by bit he collected his thoughts and put them in order, separating dreams from reality, where he placed all the events of the day before, though in fact they hardly seemed real at all.

He decided not to think about it anymore. Not for the moment, at least.

M. Villata, *The Dark Arrow of Time*, Science and Fiction,
https://doi.org/10.1007/978-3-319-67486-5_3

To start the day, he strode up to the big window, ramming on his glasses as he did so. For a better look at the landscape, he went out on the balcony. The place was stupendous, as he had been able to notice yesterday evening. He promised himself that he'd give it more attention in the future.

He washed, in the Alkenian way, in a glass booth that filled with detergent steam. He put on a shower cap so as not to wet his thick wavy hair, which fell almost to his shoulders. The steam was followed by a real rinse, with real water.

He phoned Six, to find out where to have breakfast. They told him where to find the restaurant, where he ate and drank plentifully.

He went to register, hoping in his heart not to meet Kathia.

He didn't see her. They gave him all the instructions he would need to settle in at the Center, and even a diskette containing useful information for his stay. This was until he could be issued with a computer for receiving all the info online, since his portable was obsolete and no longer compatible. Right after that he passed his eye examination.

Then he went to his department, where he was expected. Here, too, they gave him a diskette with which he could bring his research 'up to speed'. He had twenty days to study it and assimilate it, more or less. That 'more or less' encouraged him a bit, since he would have to be looking at the findings from forty years of work.

For the whole day, except for lunch and supper, he buried himself in his work, promising himself a little break, maybe tomorrow, with a nice walk around the lake.

Actually, he didn't go out on the following day, and not even the morning after that. Barely time for meals, without talking to anyone, and a breath or two of fresh air on his balcony. Immersing himself in his studies was everything he could wish for, and he didn't want any distractions.

As he was going out for lunch, however, the inevitable happened.

Just like the other times, more or less instinctively, he glanced down Kathia's corridor. Usually it was empty. That day, though, he saw her coming around the corner, down at the bottom. She was looking at some printed sheets and didn't see him right away. He took a step back, concealing himself behind the corner, uncertain whether to turn around and go back. Too late, here she was with a big smile, as if she had been expecting to find him there, behind the corner.

"Good morning!" she said, winking.

Actually, she wasn't winking, but she continued to open and close her eyes.

"Are you having trouble with your eyes?"

"It's these contact lenses they've given me, they should fit better. They must have got the wrong size."

"I've not put mine on yet. It takes me a while to get used to new things."

"So you're still putting on your glasses to go out?"

"Well, I hardly ever go out. The lenses would certainly be convenient, though. Do they really work as well outside as in here, with the white light?"

"So it would seem. If only they didn't bother me so much.... Were you going to lunch?"

Helias nodded.

"If you'll wait a sec, I'll come too. I'm just going to take off these lenses, I can't stand them anymore."

Well, he could hardly refuse, could he?

Kathia returned a moment later, clearly relieved, even if her eyes were still watering.

They served themselves at the buffet and sat at a table, facing each other.

"So, how are you doing here? I like it well enough."

"So do I. Or should I say, 'I acknowledge that'?"

"Touched a nerve somewhere, have I?" she said with her usual disarming smile.

He didn't answer. He was looking at her tray: he was appalled by the amount of food on it.

"Is that all for you? Or are you hiding somebody in your room?"

Still chewing, she nodded, then shook her head.

"I'm quite a glutton. Don't worry much about my figure."

She was still wearing loose coveralls, though not the same ones.

"I wouldn't have said so, you look like you're in pretty good shape to me."

"Thanks", she said, sipping a little water. Adding, "How are you doing with the voice commands?"

"I haven't tried them yet. As I told you, I don't trust new things. I always use ordinary things, even the intercom."

"You're old-fashioned, good for you. I'm making a real mess. Yesterday, instead of turning on the light, I turned on the microwave. And instead of switching on the computer, I opened the window. Not to mention the filter panes, I haven't understood a thing yet."

"It's easy. There are four buttons. The first on the left for completely transparent. The second works through pressure, the longer you hold it down, the darker the glass gets. The third lets you see out but not in. The fourth is for a 'Venetian blinds' effect, with transparent and semi-transparent strips. Obviously, combining them together, you can get different effects, if you like, different on different glazing panels, different windows."

"And to think that with the voice commands I barely managed to get light and dark…."

"Probably the fault of the Swedish accent."

She laughed, and regarded him for a long time, still eating.

They continued to eat in silence. Helias with his eyes on the tray.

"Tomorrow is a day of rest." she said.

"I haven't even looked at the calendar yet. What is it, a sort of Sunday?"

"More or less. It's a pause for socializing, or for relaxing at least. It's one out of every five days."

"Ah. I've studied like mad for two whole days. And today makes three. I'd say I almost deserve it."

"I'm going to the lake."

He was about to answer, "I'm not", but he thought better of it.

"And you?"

"I don't know yet, I was thinking of going round."

"On foot? But of course, you're old-fashioned."

They finished lunch and rose together.

"Why don't you come to the lake with me?"

"To do what? To have you tease me again?"

She was quiet for a moment, as they climbed the stairs.

"You didn't like it?"

"No."

"And if I were to tell you the truth?"

"I don't trust you."

"You'll come?"

"Maybe."

They arrived at her corridor.

"You'll tell me the truth?"

"Maybe."

They laughed. She held out her hand and he put his close to it. The same sensation of warmth. Like a promise.

Helias went back to studying, for the whole afternoon.

Before supper, he decided to go for a stroll. He walked along the lakeshore in front of the castle, looking around. The castle was immense. It occupied most of the ground between the mountain and the lake. In fact, it was almost embedded in the mountain, down there at the northwest corner.

Two sides of the castle gave onto a flat stretch of land. One faced roughly east, toward the lake, where he was now and where, luckily, his room was located, on the right, or in other words toward the north. The other side faced south, onto a plain that further on rose gently toward the mountain.

He went back in and had supper.

Then he stopped by the emporium, where he had all the possible filters for his camera lens explained to him. He chose the one that would have given him the same colors he saw with his glasses, and an adaptor ring.

He went up to his room and put the filter on the camera. He went out on the balcony and looked through the viewfinder: splendid. He snapped a few of the best shots, even though, by now, the lake was lit only by the last shimmer of the sun setting behind the mountain.

He went out with the camera, to try some more shots. He went around the east and south sides. When he reached the west side, he had to start climbing up a trail between the rocks. The castle was literally embedded in the mountain. One slip, and he would have ended up against the top of the surrounding wall. At a certain point, a sheer cliff ahead of him prevented further progress, and he was forced to descend toward the wall. He followed the wall as far as the northwest corner, along a sort of flagged path. Here the mountain was less rugged, with a few grassy patches. He went down along a crest that flattened out as it approached the lake. It was an artificial crest, in the sense that here the mountain had been excavated to make room for the north wall, dotted with windows and balconies, like the walls on the east and south. Most of them were probably living quarters, while the majority of the offices and labs gave onto the inner courtyard. Many of the windows were dark, perhaps just screened or obscured from inside. Others were lit but milky, giving no view of the interior. Others, a very few, were transparent, showing details of the occupants' lives.

Outside, the light was fading fast, even wearing concentrator lenses. The building's external lighting went on, yellow and orange, just as he had seen that first evening from the spaceship, approaching the north side.

There was a meadow between the crest he was walking along and the castle, probably artificial, sloping slightly and punctuated with shrubbery. He preferred to go that way, safer in the dark, and made his way diagonally across the meadow. He paused for a couple of snapshots with the yellow and orange lights. Through the viewfinder, he saw a light go on and a girl coming toward the window. It was Kathia.

She fiddled a bit with the window controls, making several attempts. At the end, she settled for the Venetian blind effect. Without perhaps realizing that you could see in from outside. And in fact she began to undress.

Helias, embarrassed, hurried to take his picture before the worst could happen.

He was about to put the camera away and leave when the girl, now in her panties, came up to the window and put on her glasses, as if she had heard something.

Helias froze, crouching near a bush and, to understand the girl's movements, couldn't help but look at her.

She was just beautiful, he thought. Now she was looking in his direction, and he even held his breath. He was still doing so when she moved away from the window, but this time for other reasons: the girl had taken off her panties and was going toward the bathroom.

It was the right time to leave, like a true gentleman.

No, he couldn't refuse a heaven-sent gift like that.

She left the bathroom door open. Quite rightly.

From where he was, he could see half the shower stall.

He shifted position. And used the camera's zoom lens, for the details.

No, the steam shower, no. The rinse was much better.

Almost without realizing it, he started filming a video, feeling not the slightest embarrassment. The gentle water rinse. The massage shower, with dozens of jets from all angles. Better than any advertising footage on the subject. Thanks, obviously, to the leading lady. The blow-dry, with the blonde hair floating freely in the warm air, while the base of the booth and that sublime form rotated gently.

Helias let himself take a breath for a moment, while she came out of the shower and disappeared into the part of the bathroom he couldn't see. After a few minutes she reappeared with a new pair of panties and sat on the edge of the round bed. She sat there for a while, plucking at her lower lip, her elbows resting on her thighs. Then she laid down and turned off the light.

Helias, shaken, stretched out on the grass to look at the stars.

With his elbows on the desk and head between his hands, the professor said, "Are you sure he's up to it? It's not a trivial thing, you know?"

"We're almost certain. In any case, we've got to take the risk, we don't have much time."

After pronouncing these words, Mattheus walked up to the window, which looked out on the inner courtyard where Nasymil cast its frigid light, contrasting with the orange lamps.

"How do you propose to convince him?"

"We'll play the cards we've been dealt. Though it shouldn't be necessary."

The next morning, Helias went down to the lake. There appeared to be no sign of Kathia.

To the right, the lakeshore was perfectly flat, but she wasn't there, at least among the people nearest to him. At the left, the shore was rocky, it was easy to image that there were hidden beaches and coves.

He walked up to last night's crest, to have a better view.

Among the rocks a bit further down, he thought he caught a glimpse of a blonde head.

He came closer. It was she. Half reclining on a rock, enjoying the warm morning sun, wearing a bikini bottom and a loose short top with shoulder straps. He continued to approach, in silence.

"Come on, I was waiting for you."

Maybe she had seen him coming out of the corner of her eye, through the dark glasses.

"How did you manage to hear me? I didn't make any noise."

"I can feel you, when you're near."

"What do you mean?"

She had sat up in the meantime, knees bent and arms around them, eyes forward.

"I feel your presence, not because I can hear you or because you have a particular smell. Sometimes I even know what you're thinking…. That's the way we are."

"We? Who? You and Mattheus?"

"Not only."

He had sat down next to her, fully concentrating on her words.

"Explain yourself better."

"It's a very long story."

"It doesn't matter."

"It does matter, though. Because I'm all sweaty and this is the best time of day for a nice swim. Why don't you come too?"

"But isn't the water cold?"

"No, it should be wonderful now."

Helias kicked off his loafers and dipped his foot in the water.

"But it's freezing!"

"Don't be silly, it's great."

She was already in the water, which reached halfway up her thighs. A few more steps and then she glided under the water, already deep here.

"But where's she from, that one, the North Pole?" mused Helias.

She surfaced.

"I'm Swedish, remember?"

"Rubbish. You don't have a Swedish accent."

"Touché!"

And then back under water. Cavorting, splashing and doing somersaults, occasionally coming up, head and shoulders out of the water, and then plunging back down.

Helias watched her, enchanted and a bit envious.

Then she emerged from the water, dripping all over, her thin top almost transparent now and clinging to her breasts.

"How do you manage to see? Are you wearing the contact lenses?"

She shook her head.

"I'll explain later. I haven't finished yet."

She clambered up a rock overhanging the deep water.

Understanding her intentions, Helias pulled out his camera. He filmed the dive and the subsequent exit from the water, mirroring all the colors of the surrounding rocks, in a beautiful glancing light. And a strange reflection in her eyes.

"Ah, the famous camera."

"If it bothers you, I can delete the whole thing."

"Of course not, why would you want to do that?"

"Well, you know, privacy...."

"That's nothing compared to violating the privacy of your thoughts."

"Why did you say 'the famous camera'? Did you 'feel' me last night?"

"Yes."

"I was about to tell you. And apologize. Or at least ask permission."

"You're lying."

She was right. She dried her face with the towel and put her glasses back on.

Then she propped herself on her elbows to dry off in the sun, breathing a little heavily and, in her semitransparent top, visibly chilled.

It was too much. Helias, sitting next to her, had to look away.

"Are you angry?"

"Not at all. But, please, don't lie again."

"You'd have every right to be angry. I took advantage of your inexperience with the windows."

"Go ahead, say I'm just hopeless. I thought I had fixed it so people couldn't see in. I realized almost immediately, though, when I 'felt' you."

"Why didn't you darken it?"

"I was afraid I'd mess things up even more."

They laughed.

"You could have pulled the curtains, the real ones."

"Actually, I didn't want to disappoint you."

She said it in an undertone, as if she were talking to herself.

She stood up, took the towel and handed it to him.

"Would you hold it for me, please?"

He looked at her questioningly.

"Like this." she said, spreading out her arms and moving toward the rock.

He held out the towel. She took off her top, still wet, and hung it on the taut edge of the towel. Then the bikini bottom. Same manoeuvre.

"Could you hold them for me, please?"

"I'd like to oblige, but, believe me, I've only got two hands."

They laughed again. And looked at each other over the top of the towel.

She spread the two garments on the rock, took the towel and wrapped it around her.

"Shall we go?" she asked.

"And the explanations?"

"You're right. But not yet. First I have to show you something. Coming?"

Maybe she was teasing him again. But he had no choice. And they started back toward the castle together.

"What is it?"

"You'll see."

In silence they climbed the stairs and passed through the corridors, reaching her room, where whatever she wanted to show him probably was.

"Wait a second."

She went in and, pushing the door behind her, walked toward the bathroom.

The door did not swing completely shut, staying open by considerably more than a crack.

She clearly was going to have a shower. He saw the towel drop to the floor. He moved away from the door and leaned on the wall. Another shower might be fatal for him. In any case, and fortunately for him, the opening didn't provide a view of the bathroom.

He waited a few minutes, while wave after wave of questions came to mind.

All of a sudden, he heard her voice saying, "Come here."

He approached the door, but the crack had gone dark. Then the door flew open and a hand grabbed his and pulled him in, forcefully. The door closed behind him.

In the dark, his hand was dragged behind her back, covered in a thin bathrobe. Before he knew what was happening, their bodies were in close contact, with an intense shiver. And she kissed him, deeply.

Then a hand snaked around his body. And he had the feeling that he wouldn't have forgotten that contact, even in a thousand years.

As it happened, he had already forgotten it after two hours. Two hours spent all over her round bed and its immediate surroundings. Exploring all the points of the compass and the four quarters of the heavens. In reality, that first fleeting contact was submerged by a lengthy sequence of soft, warm embraces. Now gentle, now fierce. Along the bastions and the turrets, in the echoing halls and the narrow passageways, through to the most secret chambers in the vanquishable strongholds.

Chapter 4
Why Did You Do that?

"Why did you do that?"

They were the first words that Helias pronounced, in the semi-darkened room, returning from that dizzying excursion.

"Shouldn't I have? Are you sorry I did?"

"No. I didn't mean that. You're wonderful. It was one of the greatest things that ever happened to me, at least lately. I was just wondering how it happened, seeing that we've known each other for so little time. I don't know anything about you."

"You're right. It's just me that knows you well."

"Yes, I know you 'hear' me, that you read what's going on in my mind. But only for the last few days. How can you say you know me well?"

Kathia, lying alongside him, took his hand and was quiet for a while.

"Not only for a few days. Almost three months by now. I think I know you better than anyone else, more than a wife knows her husband after a life spent together."

"You mean you already knew me, and 'felt' me on Earth, three months ago?"

"Yes."

"I don't remember you. True, your face is familiar, but I can't place you in any clear memory. That's why I thought I had already seen you before."

"I was around. At the university dorm I had the room above yours, and I listened to you there. When we happened to cross paths I made sure you didn't notice me. I would sometimes sit behind you in the dining hall, but there was too much 'interference' and I couldn't feel you very well."

"But why? Why this sort of mental stalking?"

"I, or we, rather, needed to get to know all about you."

"You and Mattheus?"

"Yes. Not only."

"Why did you need to know all about me? What am I for you?"

"It's a long story. We need you, Helias."

Disoriented and irritated, Helias moved to get up, it was all so ridiculous.

But she squeezed his hand again, and bent forward, seeking his lips.

M. Villata, *The Dark Arrow of Time*, Science and Fiction,
https://doi.org/10.1007/978-3-319-67486-5_4

And once again he felt the soft breast brush against him, and press against his chest. And a warm inviting hand.

"I want you to know that what happened today has nothing to do with all the rest of it. I looked inside you, Helias, and I saw so many beautiful things, better than in so many other people. And I wanted you to know me too, to come inside me. I think I love you. And that should answer your first question. Other answers will follow. Later."

An Alkenian shower together was a temptation they were unable to resist.

He was enthusiastic. Even too enthusiastic. So much so that she decided to put a stop to it with a cold water rinse. Otherwise they would have started all over again.

Lunch took place mostly in silence. Too many ears in the vicinity. He formulated questions in his mind. She thought the answers.

"I noticed your accent has improved a lot lately. The vocabulary too."

"Since you brought it to my attention, I've practiced a lot. On Earth, I practically only talked to Mattheus and I didn't know I was so bad."

"How long had you been on Earth?"

"A little more than three months."

"Just the time needed to…."

"Just the time needed."

They were almost whispering, but then they realized that it made things look even more suspicious. They fell silent, and didn't speak again until they left the room.

"Mattheus doesn't have your accent."

"He's traveled a lot."

"Were you born on Earth or Alkenia?"

"Neither."

It seemed that surprises would never end.

"But there aren't any other planets that…. Except for Murya, but that's over on the other side, more than fifteen parsecs from here."

"As far as you know."

"That are known about. Certainly. In fact, it seems that down on Earth nothing at all is known. Fault of the distance, partly. Certainly."

They arrived at the lakeshore. To avoid the swarm of people on the beach, they walked toward the rocky area, where they had met a few hours before. They took a broad path that started where the artificial crest flattened out near the lake, and appeared to wind through the rocks on the north shore.

They crossed a stream that flowed into the lake, ten or fifteen meters further down. Kathia took off her shoes so as not to wet them. Helias tried to balance on the stones that rose above the water, but slipped and ended up soaking his beach loafers, promising himself to wear more appropriate shoes next time.

They walked for a while in a nearby meadow to dry their feet, and then went back to the path.

"My people left Earth almost a century and a half ago, when the transmission technique didn't exist yet, except in theory. There were seventy of them, almost all of them hibernated, bound for Alkenia. Halfway there they hit a space-time warp caused by a little black hole that wasn't marked on the charts. They were thrown back by nearly five centuries, as the astronomers on board were able to calculate when they woke up. And a few parsecs out of their original route. They thus discovered a solar system that nobody had known about, but with a planet that was decidedly hospitable, better than Alkenia. They had no reason not to stay there, especially since it was impossible to regain contact with Earth. They were supposed to have been the first explorers of Alkenia, which at that time was known only through the fly-by probes."

She paused, while Helias put his loafers back on, though they were still damp.

"Go on, please."

"We settled on that planet and, over the centuries, developed our science and our technology, as well as our philosophy, five hundred years ahead of Earth. We could have gone back to Earth and thumbed our noses at our ancestors. But we decided, then and for ever, that we would never interfere with the past, as one of our civilization's first ethical rules."

"How could you have gone back to Earth, if there's no trace of your return in Earth's past?"

"In fact we didn't go back."

"But suppose you had decided to, knowing full well that there is no record."

"Basically there are two theories. The first maintains simply that the return took place but didn't end up in the history books. This theory thus denies that the course of events could be changed, or rather, states that the course of events changed exactly the way it happened. So no parallel futures, or anything like that. The second theory, on the other hand, leaves all possibilities open. A time traveler could destroy half the world with a bomb or something, or interrupt his own bloodline and so prevent his own future birth, none of it matters, it will result in a history that's different from the one he knows, while the latter will continue unimpeded elsewhere. In this case, as you can see, if one of us had gone back and messed things up, it would have 'opened' a parallel future that has nothing to do with us."

"So according to this theory, the history of the Earth as we know it didn't receive visits from the future, at least not evident ones. I've also heard about a theory of probability waves...."

"Yes, but it's not very popular with us anymore."

"I interrupted you. Please, go on."

Kathia had sat on a rock jutting out over the lake and was looking down, lost in thought.

"You're not thinking of diving from here?"

"Why not? It's a bit high, but it can be done. Too bad I don't have my swimming things."

"I think you're out of your mind. Don't even think about it."

"It's a temptation. I could go skinny-dipping, there's nobody around, and then dry off in the sun."

"Don't, please."

"Why?"

"Because I don't want to have to scrape you up with a spoon. And then…."

"And then what?"

"And then I didn't even bring my camera. I wouldn't want to miss something like that for anything in the world."

"Something like what? You scraping me up with a spoon?"

They laughed, and looked at each other for long time. Then Kathia took off his glasses.

"I think I love you, Helias."

"Me too."

And in the pink dusk he took off her glasses and sought her lips.

Smiling, he put her glasses back on her smiling face. Smiling, she did the same for him.

"I have to show you something, Helias."

"Again? In your room, mean?"

"No, here."

She took off her glasses again and turned away. He saw her rummaging in her pockets and pull out a little case. She bent her head forward until her eyes almost touched the open case.

"What are you doing?"

"Just a second."

First one eye and then the other. She closed the container and put it away.

"I took off my contact lenses."

"You were wearing your lenses? With the glasses?"

"No, not the lenses they gave me here. I don't wear them, they bother me, as I told you. I took off my own, the ones I wear all the time, or almost."

"Are you nearsighted? Back on Earth, people have an operation for that, usually."

"No, I see fine. We've eliminated all eye defects genetically."

"I don't understand then. Why do you wear contacts?"

"So as not to show my real eyes."

Helias waited, in silence.

"Are you ready?"

He nodded, with a sort of moan.

Kathia turned, and there was that strange light in her eyes again, like that morning, when she emerged from the water.

Kathia came closer, so that he could see better.

Her eyes were blue, the lightest of light blues, crisscrossed with tiny multicolored streaks, from yellow to orange, from green to cobalt. They seemed almost phosphorescent. Helias had never seen anything like it, nor imagined such a thing could exist. His first reaction was one of fear, almost. In fact, his hair stood up. It wasn't the Kathia he knew, and an alien glance has something inimical about it, triggering an instinctive, ancestral fear.

She noticed and murmured, "Don't be afraid, it's still me."

He continued to look at her, asking himself how much the glasses altered the colors.

"No, they're really like that, or almost." she said, anticipating his question.

Bit by bit he got used to them. More than used to them: he thought they were beautiful. Absolutely beautiful.

"If you like, I'll put the lenses back on. Or the glasses."

He shook his head. He couldn't stop looking.

"You didn't have them this morning at the lake, either. You had your eyes half closed and I never looked at them close up. But at a certain point I noticed a strange glimmer, while I was filming you. And so your eyes are adapted to this light, you don't need aids of any kind."

"Actually they're adapted to any kind of light. All they need is a handful of photons at any frequency between the far infrared and the near ultraviolet. They're the result of some sophisticated genetics. First we modified them to adapt them to the light of our planet, and then we perfected them for any occasion. They adapt automatically. I can add Alkenian contact lenses or glasses, it doesn't change much."

"But you don't want anybody to know, and so you do what everybody else does."

"Right. Nobody must know that our race exists. Except for you and a few others."

"Why?"

"As I was telling you, our civilization is hundreds of years ahead of Earth. Not so much because of technology or genetic engineering, which we use very sparingly, despite appearances, but because of our psychic qualities and even more so because of our philosophy of life, and the ethics that derives from it. In a way, we got over our initial enthusiasm for the extraordinary and almost unlimited potential offered by scientific and technological advances. We understood where the catch was. For every exciting innovation, there's always another side to the coin. That doesn't mean we decided to go back to the stone age. Far from it. We just became more prudent and developed a certain mistrust. The principle we go by is that of taking minimum action. In other words, we introduce only those innovations that we believe to be indispensable or very important. There are special councils that scrutinize every proposal. With our knowledge of biogenetics we could do practically anything, but we limit ourselves to the strictly necessary. And that brings us to the crux of your question. What would happen if our advanced knowledge, which we use, or at least believe we use, wisely and with the utmost caution, were to end up in the wrong hands, like those of the earthlings, for instance, who might be in good faith but are inexpert and primitive?"

"So why don't you just stay quietly at home on your own planet instead of gallivanting around all over, running the risk of being discovered?"

"And wait for them to discover us on their own. It's about—or rather, it was about to happen—you know? Fortunately we acted just in time."

"Meaning?"

"I'll tell you when you've calmed down. You don't have to feel wounded in your identity as an earthling…."

"Sorry to be so ingenuous. After all, I'm just a poor idiotic earthling."

"Don't be so touchy. You're doing yourself a disservice. That's just the way things are. Without insulting anybody. You of all people, you're usually so clear-headed and intuitive."

They fell silent for a while. Helias with downcast eyes.

"You're right. Sorry."

"You don't have to excuse yourself."

"My bottom is sore from sitting on this rock. Shall we continue up the path?"

"I was about to suggest it. My bottom, though much more evolved, doesn't feel any better than yours."

They rounded the north shore in silence, while Helias tried to get his head around all that unexpected and surprising information.

After a short climb up the mountainside, the path began to drop toward the eastern shore, gentler, with rolling meadows and clumps of shrubs, finally reaching the tributary river, its broad churning surface spanned by a narrow wooden and stone bridge that enabled them to continue their circuit of the lake.

"According to our rules, we couldn't visit Earth before our departure. In the meantime, more than two centuries before, we had invented transmission, or we had put it into practice, rather, starting from the theory existing on Earth, but which involved problems and difficulties that at the time were insurmountable. We patrolled the nearby planets, especially Alkenia, where we knew the earthlings would arrive one day, and where they would be able to discover us easily. We had to keep an eye on them, and so we built two bases on this planet."

"Where?"

"One's very far from here. Almost at the opposite end of the planet. The other is right in front of you, inside the mountain, behind the castle."

Helias turned toward the lake and the far shore, dominated by the castle, and looked at the mountain that in turn dominated the castle. From where they were standing, the mountain was much higher and more imposing than it looked from the opposite shore.

"And so it was you people who made sure that the Kusmiri Center was built right here. How did you do that? And why?"

"As you know, much of Alkenia's scientific work takes place here. That way we can monitor your progress from close up, it's no accident that I'm here as an archivist. Not only, but we can use our base, which we reach through the castle, without arousing suspicion. Here we pass as earthlings doing our jobs. Anywhere else, our comings and goings would have attracted unwelcome attention. How did we do it? The way we always do, ever since we started keeping watch over you. We go to Earth, assume an earthling identity, we infiltrate where we want to, in all the key points, and in practice we monitor just about everything."

"It can't be all that easy to 'invent' new earthlings, from one day to the next."

"We can breach any of your archives or computer systems and edit them, without anybody realizing what happened."

"Correct me if I'm wrong. So, a little less than twenty years before your departure from Earth, you transmitted yourselves there, in order to arrive a bit afterwards, as your rules require, and you started pulling all the strings, like puppeteers…."

"Almost. You're pretty close. In reality, though, things are a bit more complicated."

They were crossing the bridge, and the noise of the rushing river nearly covered their voices. They stopped to admire the view, and once again Helias regretted not having brought his camera. A breeze had come up, ruffling their hair and the water, breaking the surface of the lake into myriads of tiny reflected suns that rose and fell in the tossing waves.

They walked hand in hand, like a couple of sweethearts anywhere, and continued toward the south shore, with its welcoming meadows and small yellow and brown trees, where the lake found its natural outlet and another bridge led back to the castle.

"What's your planet called?"

"Thaýma. Which in ancient Greek means 'marvel'. A naive name, but that's how it appeared to our ancestors. You should see it."

"I'm counting on it."

"I'm glad you're here with us, Helias."

"I just hope I don't disappoint you. I'm anxious to know what you expect from me, but I see you'll have to give me a lot more information before getting to the point. But what makes you think that I won't refuse, or that I won't betray you? Or are you thinking of bumping me off if I do?"

"I know you're joking. Do you think I'm capable of killing somebody? You, especially?"

"Well, maybe not you. But what do I know about the others?"

"We're nonviolent. Inasmuch as possible. In any case, human life is sacred for us too."

"But you're sure I'll accept?"

"Practically yes. I know you well, for three months by now. I'd be an idiot if I got everything wrong."

"Tell me more."

"To intervene on Earth, and thus cancel all traces of our planet's existence, we had to use transmission. Transmission as you know it makes it possible to zero the distance between the two planets and the people being transmitted are unaware of the passage of time, but if you want to go back to where you came from, or even just send a message, it would get there forty years later. Hardly convenient."

"'There's the trick' Mattheus told me, and he added that I'd know about it 'in due time'."

Kathia had that look again, that sober, respectable expression she had mocked him with the first evening. To get even, Helias tried to pinch her, but she ran off. He went after her and, tripping, brought them both down, rolling in the grass. Then he made a dive and kissed her.

"You're a monster" she said, smiling.

"Listen to who's talking. The abominable spacewoman from the unknown planet."

"Oh, really? Not too abominable for you this morning, I'd say."

"Because you hadn't shown me those horrible eyes yet."

And he slipped a hand under her light coveralls.

"You're just a horrible monster. Always ready with those hands. And with something else too...."

They heard voices approaching and they hastened to regain their composure.

A group of people appeared from behind the rise. Probably back from an excursion in the area. It was a holiday and the weather was magnificent. Nevertheless, Helias was amazed at how few people ventured far from the Center. Maybe it was that strange, weak light that put them off. And the idea of being on a planet part of which was still unknown, where you don't know what the trees and the flowers are called, and the grass is thick and rough. Where everything takes on an unusual color, the color of dreams and not of reality.

And yet many of the inhabitants must have been here for a good few years. But apparently they still weren't used to it. Or they were simply lazy and incurious. And they weren't photography buffs.

Kathia, ever cautious, put her glasses back on. They let the group pass and started off again.

"For transmission, you need to have a receiving station, similar to the transmitting station. Both have an antimatter ring, a few dozen meters in diameter, held together and protected by an intense magnetic field. You already know these things?"

"Very vaguely, yes."

"At the time of transmission, the craft, or rather the spaceship, is launched into the ring and at the same time the magnetic field is released for a few moments. The spaceship is incorporated in an equal mass of antimatter. The result, as you know from elementary physics, is an electromagnetic wave that propagates at the speed of light. But it has to know where to go. So there has to be a receiving station, ready to separate the two components again. One of the most critical points is precisely that of maintaining the identity of the two parts. The theory was known on Earth, but they were way behind in solving the technical problems. To make a long story short, we went to the appointment on Earth bringing with us a good stock of antimatter, which they didn't know how to harness yet on Earth. We infiltrated the research institutes of the day and, taking the necessary time, we 'reinvented' all the technical details that were needed. In the meantime, Alkenia had been colonized and, starting from Earth, we infiltrated this planet too, to 'help' organize the receiving station, among other things."

"There's something that escapes me here. Why help the earthlings with the transmission technique? Didn't that also increase the risk of being discovered?"

"No, far from it. They would have traveled anyway, even without transmission. With transmission, on the other hand, we had an enormous advantage, but you don't have what you need in order to understand that yet."

"The advantage is the 'trick'?"

"Bravo! Yes. With normal transmission trips are, yes, faster, but the time that goes by between the outward and homeward bound legs are still unacceptably long to permit repeated and efficient action, as you earthlings know perfectly well, as it takes you decades to exchange information. We had the key, or in other words the 'trick'. We could practically reduce round-trip times to zero, in a way you would never have been able to figure out on your own. Mattheus and I didn't leave here forty years ago to go to Earth, but only a few months ago. Once the two stations were ready and interplanetary voyages were replaced by transmission, we could also roll out our 'back-now' technique. We could go back and forth as if the distance was really nothing. From the Earth to here, twenty years went by, same as for you, but we used back-now from here to Earth, and that way we got the twenty years back. In theory, we could do the whole trip in one day, as if we were going between two adjacent planets. It didn't matter if what was regarded as 'now' in the two planets happened to be twenty years apart. Only our first people, the ones who 'prepared' the stations, had to actually experience the real distance. They went to Earth normally, leaving forty years before they could have returned. And those of us who stayed here heard from them only after forty years."

"Once the stations were ready, couldn't the back-now technique be used from there to here?"

"No. You can only do it in one direction. Otherwise we would break our rule of not interfering with the past. We could have left from here, gone to Earth, and come back here forty years ago. That wouldn't have been good."

They were crossing the second bridge and were nearly to the end of their circuit of the lake. They could already see the swimmers on the west shore, below the castle. There were more of them than there had been a few hours earlier. Now the water must be warm, even for him.

Neither of them wanted to mix in with that mob, or go back to the castle yet. They sat down in a meadow again. Kathia had picked up a few sticks and she arranged them in front of herself.

"See? These two vertical sticks, parallel to each other, represent the histories of the two planets. The one on the left is the Earth's time axis, which runs from the bottom upwards, while the one on the right is Alkenia's. My ancestors left at this point of the Earth axis, heading for Alkenia. Then they were thrown back in time and ended up here, on Thaýma, on this third stick near Alkenia, much lower down. Here they waited for the flow of time to bring them within range of Earth, a few years after their departure, and they went there, diagonally, along this blade of grass, while around twenty years went by on the planets. Ten years or so later the transmissions began, and earthlings and Thaymites could reach Alkenia along these blades of grass, diagonally from the bottom to the top. At this point, however, the Thaymites can also make the trip back, along the same blades, but from the top to the bottom. And subsequent history is a series of parallel diagonal blades."

Helias picked a blade of grass of his own.

"And this? This never happened?" he asked, placing the blade in the opposite direction, from Alkenia to Earth.

"This is what happens normally for information and exchanges of material. For human lives, it's too risky. Nobody can guarantee that the receiving station on Earth is working, after forty years. It's prohibited. We can't go to the future, in a spaceship, and then, if everything's working, return with back-now and say 'okay, go ahead', or even just send a message to confirm. Because that way we'd know the Earth's future and we couldn't interfere with its past, or in other words, we could no longer do anything now."

"How did you manage the first time with Alkenia? How did you know that the station here was okay?"

"The transmission of information, meaning electromagnetic signals, is a trivial problem, it doesn't need antimatter and it was ready almost immediately. For that too, there's a sort of back-now, and so we were in continual contact."

"But the earthlings didn't know about it. How were you able to convince them?"

"Well, even then there were simulations of what was going on Alkenia. When the probability that the station wasn't ready yet dropped below two percent, some of us volunteered, knowing that the risk was almost null. Then the probability kept dropping and by now the ice was broken, and so the earthlings started volunteering too, and finally news arrived that everything had gone as expected."

"Just as well they didn't know about 'your' black hole, which doubtless wasn't included in the simulations. Otherwise they'd have been a bit less trusting."

Kathia kept looking at the time.

"Do you have an appointment?"

"Yes. Excuse me a moment."

And she pulled out her 'cell'. She, too, spoke in a strange dialect, like Mattheus on the spaceship.

"I've got to go, will you come with me as far as the castle?"

Helias, rising, looked at the mountain.

"Yes, there." she confirmed. "There's a meeting."

They started off.

"When will I know the rest? When will I see you again?"

"I don't think this evening. We usually finish late. Tomorrow morning I've got an appointment with the optometrist, for this business of the contact lenses that bother me. Right afterwards? Shall I get in touch with you myself?"

"Yes, but anyway give me your code."

"'olsson', with two s's, space 'k'."

"That's your surname?"

"Yes, the Earth one."

"Why do you change name?"

"We do it when it's advisable. In my case it's not to reveal a connection that might cause comment."

"Which is?"

"My real surname is Bodieur."

"You're relatives?"

"He's my uncle. Wait, I'd better put my contacts back on. Cover me, please."

She pulled out the case and leaned on Helias, whose back was turned to the swimmers.

"Is anybody coming from that direction?"

"No."

She bent over the open case. He couldn't resist kissing her hair. She raised her head and smiled at him, with her normal blue eyes.

"Isn't there a danger that the optometrist will notice?"

"No. They camouflage perfectly. It would take one of ours, who knows what to look for.... Here comes somebody."

"Kadler? Helias Kadler?"

Helias turned.

"Yes?"

"You know, I can recognize you even from behind.... That unmistakable mane of yours...."

Where had he already heard that irritating voice? And who was this who was addressing him so familiarly?

But of course, the eternal pain in the ass. That person you'd never want to meet again. But instead you're always running into him, even in the places where you'd least expect it. And above all in the worst possible moments.

Kathia, who 'felt' everything, smiled at Helias's discomfiture.

"Yeah, it's me. We did our doctorate together, remember? Geremy Stuerz."

And he put on a disappointed look, since Helias, who wanted nothing more than to pretend he was someone else, looked at him questioningly, his eyebrows raised.

In reality, he was asking himself what sort of curse had befallen him, that such a radiant day should also have to end with such an unwanted encounter.

Kathia watched him maternally, and smiled again.

He glanced at her reproachfully.

"C'mon, you can't possibly not remember me. We defended our dissertations together...."

He was one of those people who you try to wipe out of your memory. Clearly without success. Because you end up face to face with them again. It wasn't the guy's fault. He tried to get people to like him. Tried too hard. He was a phony, a liar. And he had always been a sycophant, with everybody, but especially with the professors. Once Helias had caught him badmouthing him with the director. He'd realized it because they were looking at him from a distance and, as he drew nearer, they fell silent and Geremy had dropped his eyes without even saying hello.

Kathia's expression sobered, and she turned toward the newcomer.

Helias managed a slightly strangled, and none too cordial, "How's it going?"

"Hey! But I've already seen her too...."

"Yes, she was hanging around there too, toward the end...." answered Helias on her behalf, hoping to spare her an unwelcome involvement.

Unwillingly, he extended his hand, which found itself in a cold and bonelessly limp grip.

Geremy continued to talk, with nobody listening. Helias and Kathia looked at each other. She murmured, "I've got to go."

He thought "Wait another moment. I'll come with you."

"I can't." she mouthed.

"See you later, I have to go now." she said to both of them, and walked off.

Then she turned again to look at Helias, with a smile. She waved and blew a kiss. Then she looked at Geremy, whose back was to her, and her face grew serious.

He was about to run after her, the hell with this interloper.

But he wasn't able to. The other continued blathering at him. And Kathia was already far away.

That's why he detested him. Because his stupid chatter covered you like a coat of slime. He stuck to you, without even giving you time to react. And you could never shake him off. Like a tick. That you can get rid of only if you do him some damage. And Helias wasn't cruel enough.

It was one of the worst half hours of his entire life. Geremy also introduced his friends, all pretty much cast in the same mold, who were with him on the beach. At a certain point, he even made some appreciative remarks about the blond girlfriend. And it was all Helias could do not to punch him in the nose. When he was really and truly fed up with nonsense, liberally laced with memories and mentions of the good old days spent together back on Earth, which Helias, in reality and fortunately for him, remembered hardly at all, he cut loose, claiming, lamely, that he had an appointment. Just in time, since they were about to invite him to supper, and there was no way he was going to put up with torture like that.

To avoid other unpleasant encounters, he took his supper up to his room. And he spent the entire evening thinking about Kathia and everything she had told him. And above all, he kept thinking of how she had looked as she walked away, turning and smiling at him and sending him a last kiss, like a promise to meet again soon. Then her face grew serious. And Helias slept.

He dreamt of Martians. Who were going back and forth in time, running in equilibrium along thin elastic cords, stretched nearly to the breaking point and ready to snap at each step. He was a Martian too. And he jumped with the others, higher and higher, ever higher. But he didn't know where he was going, whether there was a tent atop that circus, that sooner or later would have stopped him, that he could catch hold of. Every jump brought him a little higher, and his fear grew, not on the way down, but when he realized, with each jump, that he had gone higher than the time before. It was the unknown. All the still-unanswered questions.

Chapter 5
The Next Morning It Rained

The next morning it rained. Poured, in fact. It had started around six, with big isolated drops that spattered sharply against the balcony and windows with a dry rattle that had woken him. Like the edge of a storm, but without thunder and lightning. The low, dark clouds cut the surrounding mountains in half, lopping off the tops. Then the rain intensified, veiling the opposite shore of the lake, and soon hiding everything farther than a few dozen meters away. Now, though, the rain had dropped to a steady drumming and the nearer waters of the lake were visible again, their surface pocked by the falling drops.

After breakfast, Helias looked at the weather forecasts on his brand-new computer. He discovered that the weather there, at least in that period of the year, was highly variable, and that you could expect a downpour at any moment, even when the sky seemed clear. He wondered if that was the reason there had been so little going on around the lake the day before. He went to look at the recreation and leisure page, and found that there were a number of rooms offering various opportunities for sports and entertainment in the castle, including a pool and two gyms, in addition to the sports ground on the south side that he had already seen. Also, on one out of every two holidays a plane holding one hundred and fifty passengers took off for the capital at seven in the morning, returning at ten the following night. People who chose this opportunity for a break in the routine waived their rights to the next holiday, or in other words took two out of ten days off rather than one out of every five. This is what had happened the previous day, and explained, at least to some extent, why there had been so few excursionists, as well as the poor turnout for breakfast that 'Monday' morning.

It was almost nine thirty by now. Helias had no idea what time Kathia had her appointment with the optometrist. He looked at the timetables: from nine to twelve. He glanced at the index of the material he had to study, but had absolutely no desire to plunge back into work, after everything that had happened the previous day, and especially in this brief period before meeting Kathia again. At nine forty-two, he hooked the cell to his ear, something he never did, not even on Earth, and went out. Kathia had said she'd get in touch, but she hadn't said how. Doubtless, as an

M. Villata, *The Dark Arrow of Time*, Science and Fiction,
https://doi.org/10.1007/978-3-319-67486-5_5

archivist, she had access almost everywhere, even to e-mail addresses and cell codes.

He wandered down to the optometrist's office, to see what was going on. There were two people in the waiting room. Maybe Kathia was inside. No, the office door opened and a big boy with a cheerful face came out.

He got an umbrella from the porters and went to walk along the lakeshore, to have a closer look at the evanescent bubbles busting from the surface as the heavy drops pounded the water. At ten twenty-one, after making sure his cell was working, he went back inside. He passed by the optometrist's office again. There was nobody left in the waiting room, no voices audible through the half-closed door. Before returning to his room, he turned into Kathia's corridor and stopped in front of her door, undecided whether to knock. No, better to call later, he didn't want to harass her, or look too anxious. But listened at the door: silence.

Back in his room, he checked his e-mails. Just a couple of not very interesting announcements. At a quarter to eleven he decided to call: no answer. Maybe some urgent work had arrived that she had to get done right away, or last night's meeting had forced her to change her plans for the day. But why not let him know? She could easily imagine how concerned he was. Concern that was turning into outright worry. But he didn't want to be anxious. His ex-girlfriend used to upbraid him for exactly that: "But I never told you what time it would be! And anyway if I couldn't, I couldn't. Don't breath down my neck all the time....".

Okay, okay, no panic. He was just too prone to worry, okay. It seemed to be his karma, having to wait for people. Who maybe then went away. Or, simply, didn't come back.

"Mail." said the computer's aseptic voice.

"Yes?"

"Mail from anonymous sender. Second attempt. Do you want to receive it?"

"Tell me more."

"First attempt at ten eighteen. Anonymous mail cannot be received without your permission."

"Okay, permission granted."

The message was written, no recorded sound.

"Don't try to contact me, please. See you soon, I hope."

It wasn't signed. No sender's name and no sound. Programmed to self-delete after ten seconds, leaving no trace. A few words. Written in a hurry, perhaps hidden from prying eyes. Two attempts, spaced around half an hour apart. Why not closer? Where was she? Why wasn't he supposed to look for her? Was she in danger? Why? Why all this mystery? Why didn't she tell him more, wherever she was?

He wrote the exact words on a piece of paper, which he then folded and put in his pocket. He paused while writing the last sentence: a wish and a hope. The same things he felt himself. Once again, he pictured her turning, smiling, and blowing one last kiss while she walked away. A kind of promise. Which now was reduced to a wish. A hope.

"See you soon." he repeated to himself, hoping she could read his thoughts.

He really couldn't bear to just stay there, twiddling his thumbs and waiting for who knew what. He went out, still with the cell hooked to his ear. He returned to her corridor and listened at her door. Nothing. No, a small sharp sound, followed by a sort of rustling. He came closer. Nothing else, silence. He tried to turn the door handle quietly. Nothing, it wouldn't open. Maybe his ears had deceived him. No, again a small dull noise. He went to wait just around the corner, at the intersection of the two corridors, where he had hidden two days earlier, when he had seen Kathia arriving. And once again, a face appeared suddenly from around the corner, certain of finding him there. Startled, he jumped back. It was Mattheus who, after looking over his shoulder and all around, whispered "Don't follow me. Stay well away. I'll get in touch myself. Go now.". His face was drawn, his look even more serious than usual. He had spoken almost without moving his lips, with a strange accent, worried.

So firmly had Mattheus issued his order that obeying came spontaneously for Helias. Who after a few steps cursed himself for being an idiot, for going away like that, without even asking for an explanation. But Mattheus's glance, as usual, brooked no quibbling. And so obey it was, without objection.

Back in his room, Helias tried to make some modicum of sense of the whole mysterious business. Above all, he tried to connect the latest events with what Kathia had told him the day before. He was looking for a link, a reason. Uselessly. He had too little to go on. There must be an enormous mass of things he didn't know, information he didn't have. To begin with, he was utterly in the dark about his own role in this thing. But maybe that's where the answer was. They were in trouble and needed him in some way, God only knows why. And this trouble had suddenly got worse, blowing all their plans sky-high. No, he was just trying to guess. In reality, he knew nothing, absolutely nothing. Not yet. He just had a feeling of danger, especially for Kathia, and maybe even seen in Mattheus's worried look. Or in the few, hurried words in her message. And yet, right up to yesterday, nothing looked like giving cause for alarm, not immediately, anyway.... Suddenly a chill passed through Helias. As if a photo, a single frame, had flashed by, vanishing in an instant. A subliminal message, something you are not consciously aware of, but that triggers an instinctive response. Something had happened. There had been a moment, an image, that correlated with this feeling of danger. But he couldn't remember what. And he knew that no amount of effort would bring it back to mind. Like those things that, the more you reach out for them, the faster they elude your grasp.

He went to eat. Not hungry. He saw Mattheus, at the end of the room, in the corner. From his position, he seemed able to keep watch over everything and everybody. They looked at each other. Helias saw an empty table, not far from Mattheus. He went up to the buffet. When he turned, Mattheus was still looking at him. Helias indicated the free table with his eyes. Mattheus blinked, slowly. Understood. Helias sat, trying to understand the other's intentions and noting that he had almost finished eating. Mattheus soon rose with his tray and passed close to Helias's table. A quick, almost imperceptible movement, and a fork fell at Helias's feet. Helias leaned over to pick it up, and held it out to the serious man, who thanked him. He found himself holding a tiny roll of paper that had been caught between the tines.

Not daring to put it in his pocket, he kept the roll between his fingers, his hand half-closed, until the end of the meal. He put it in his pocket when he took the ticket for coffee. Plenty of sugar. He needed it. The minute he got back to his room, he unrolled the paper and read it.

"At 14:15. Look out the window. If I look at you as I come in, take my umbrella. Otherwise follow me at a distance."

It was hand written, in tiny but clear letters. There wasn't room for another word. For the last words, he had to use a magnifying glass.

At ten past two he went to the window. It was still raining, and the sky was even darker. When he saw Mattheus arrive, Helias hit the clear button on the window, so he could be seen from outside. Mattheus, on the lakeshore, was strolling innocently with his umbrella. When he drew even with the window, he turned toward Helias, for a couple of seconds, and then returned the way he had come, slowly. Helias understood and hurried for the lobby. He arrived just as Mattheus was coming in. There were people around, and Mattheus didn't look at him. He left the umbrella and went down the corridor on his left, still without looking at him. Helias checked that no one was watching, and then followed him at a distance. Once around the corner, Mattheus had disappeared. But an elevator was going up, and stopped at the second floor. Immediately afterwards it started down again, and stopped in front of him. The doors opened. It was empty. Not quite. There was a memory card in the middle of the floor. Helias took it and stole away, like a thief.

There was writing on the card: "The password is time axis. Read offline.".

It was she! And in fact, the handwriting was different from that on the roll of paper.

To be safer, Helias used his 'old' portable. The card was compatible. He typed in 'branch'.

"Incorrect password. Make sure that your Caps Lock key is turned off and try again. You have two more attempts before the message is deleted."

It wasn't the Caps Lock. What had she called them? Twigs? No.

"stick"

Okay.

"Second password: 'Though much more evolved, has the same problems.'"

Helias was puzzled for a moment. Then he smiled. And blushed. It was definitely she.

"bottom"

Again, the message was written. No video, no sound.

"Between the strips of light you will know what you seek. Where what sees is transformed you will find it. With that you would not want to use."

"The message will be deleted in ten seconds. Nine. Eight…."

Helias had read it carefully. He wrote it down and reread it many times. And he committed it to memory. Then he threw all three messages in the toilet and flushed them down.

They didn't trust e-mail. Or the phone, or web. Or anything that could be intercepted or tapped. They knew they were being watched. Or at least that there was a probability that they were. But by whom? And why? What had happened? Why was Kathia hiding? Had they found her out?

The first message seemed intended to keep him from being involved. Or rather, to 'isolate' him from Kathia, who was probably being watched. To the point where she had to go under cover. Mattheus, on the other hand, was still in circulation, but was trying to limit interaction with him to the bare minimum. But he, Helias, still had to do his part, as this last message explained. Isolated from the others. He had to do his part without anyone being able to imagine that he was in on the game. The message was written so that only he could understand it. If, heaven forfend, it were to fall into the wrong hands. And so he had to decipher it on the basis of what only he, together with Kathia, could know. Or in other words, on the basis of what they had said to each other, or what had passed between them. Like for the passwords, easy for him, unthinkable for anyone else.

He asked himself why he should get involved in this business. That he still knew little or nothing about. But his affection for Kathia left no room for doubt. She had asked him to take action. And he would do it. Because he trusted her. Because he wanted so much to see her again. And that was the only possible route.

He had to look for something. Who knows for what purpose. He had to look for it 'between the strips of light'. No, 'between the strips of light' he would have known what to look for.

He didn't have to try to understand rationally. He had to let his intuition work freely. Intuition that shows you the answers when you're not thinking. Or at least when you're not aware that you're thinking. When you've removed all of the unwieldy obstacles that normally clutter the mind. When there is no longer anything to prevent the current from flowing along the thin wire linking the question and the answer. When the space between them is reduced to nothing. And there's no longer any difference. The question evaporates. And there is no more doubt.

Helias gently closed his eyes. Flash! It was easy.

He went to get his camera and connected it to his portable.

"Two days ago." He said. "Castle in the evening. North side." He added, since that's how he had coded the images. "Screen."

The screen appeared. And the images.

"Mono. Stereo. Holo. Mono."

He chose 'mono' so that he could increase the resolution.

The shot of the castle with the lighted window, strips of light in Venetian blind mode.

"Zoom right."

He switched to manual control, more practical to use. Zoom on the window: strips of light. You could make out Kathia taking off her clothes, but the image was overexposed, because of the dusk outside.

"Adapt. Recover."

He played around with the contrast. Nothing, too overexposed. He went ahead, slowly.

Next, the video showed Kathia in the shower, rinsing off. He would have liked to watch for a while. But this wasn't the moment. Now the exposure was adapted to the interior light and the strips' edges were blurred, almost touching each other, since there was less depth of field and the Venetian blind effect was no longer in focus. There was even less depth of field when he had zoomed in, and the result was a bright blur, where the dark strips disappeared and were replaced by soft bands of light. The 'stereo' didn't help, since the strips were almost horizontal.

When he had zoomed in as far as he could go, almost the entire frame was occupied by Kathia. When he had pulled back, he could see the whole shower stall and part of the wall behind it, a piece of the washbasin, with shelf and mirror, and a section of the partition wall that, out of focus, blocked the right hand side of the bathroom from view.

Gentle water rinse. Massage shower. Et cetera. No clue. At least apparently. Replay. Nothing, during the rinse. Then the base of the stall begins to rotate and the hot air jet starts, blowing the long hair around. Kathia's body rotated uniformly, legs slightly apart and moving her arms, to dry everywhere. Not the face. It turned along with the body when she had her back turned to the camera, or it turned to the left or the front. Then she turned her face first, before the rest of her body, as if to look at something to her left, at the right of the frame, where the washbasin was. She was looking fixedly at something. And she continued to focus on for nearly half a turn, until she had her back to him again, with the washbasin on her right. Maybe it was just a way of drying her hair better, or of getting it out of her eyes, since the main jet came from that direction. Or maybe not.

Helias looked for a frame with a wider field, one where you could see the washbasin, half concealed by the partition wall. Zoom in. Maximum resolution. Contrast. There's something on the shelf. Red. About an inch wide. Maybe a bit less. Freeze frame. No, worse, too much noise, the object is in the blurry area of the Venetian blind effect. Stack images, to suppress noise. It's flat, possibly round. It could be a computer diskette, in a round red case. But it could also be something else.

Forward. The sequence ends with the base coming to a halt and the hot air jet going off. New sequence. Kathia is out of the bathroom now and goes to sit on the edge of the bed. She seems pensive. Check the visible surroundings. Nothing in particular. She plucks at her lip with one hand. But the other hand is moving too. The forearm resting on the thigh oscillates slowly up and down, as if to avoid the bands of shadow. And there, in Kathia's fingers, the red diskette reappears, oscillating between the strips of light.

"Between the strips of light you will know what you seek...."

In the last frames, Kathia stretches out on the bed, putting the diskette under the pillow. Then darkness. And Helias switched off the computer.

It was past three thirty. The rain had thinned now, and the sky was lighter. Helias went out on the balcony and looked northward, at the lake's rocky shore, trying to remember. He closed his eyes for a few moments. He reopened them and remained motionless for almost a full minute.

He left his room. He went down to the little kitchen where people could fix themselves light meals and snacks when the restaurant was closed. Then he left the building and headed north, this time wearing proper shoes and a hooded rain jacket.

He crossed the stream, without slipping, and arrived within sight of the rock that jutted out over the lake.

"Where what sees is transformed…."

He turned and looked around, to make sure he wasn't followed or watched.

The rain was tapering off, but more black, threatening clouds were massing behind the mountain. With unexpected speed, they started dropping toward the lake, and by the time Helias was a few steps from the rock he was shrouded in thick dark fog that obscured everything over a few meters away.

Perfect. Nobody could see him. He sat on the damp rock, more or less where Kathia had been seated, and where, turning away, she had rummaged in her pockets and taken off her contacts. He tried to sit exactly where she had been. He extended his arms, feeling the surface around him in a semicircle. There was a crack in the rock, half a centimeter wide or a little more. A portion of the crack, near him, was filled with damp soil.

He, too, rummaged in his pockets and pulled out the spoon he had taken in the kitchen.

"Where what sees is transformed you will find it. With that you would not want to use."

Because he had said "I don't want to have to scrape you up with a spoon….".

He dug gently in the crack with the spoon, until he had removed all of the soil. There appeared to be nothing in the dark crack.

He slowly inserted the handle in the crack, gently working it back and forth. Nothing seemed to interfere with the movement.

At a certain point he heard a click, and felt the spoon rebound slightly in his fingers. Slowly, he pulled it out.

The little red case was magnetic, and clung to the shiny handle.

Helias was concentrating so much on what he was doing that he didn't notice that the fog had grown even darker and denser, almost as if night had fallen.

Now the fog was blowing down in gusts, and Helias took advantage of the rare moments when it thinned to mark out the way back. The fog rose and fell, but always in a crepuscular light: the layer of cloud overhead must have been very thick. The stream had swollen to a torrent and he was forced to find a place further up where he could cross, though not without difficulty.

The fog was mixed with rain now, as Helias clambered down toward the path. In the patter of the drops, he thought he heard a noise, a stone dislodged. A bit ahead of him, in the fog-bound darkness.

He froze, steadying himself with one hand on a rock, and listened hard. Nothing, only the noise of the rain. No, a rustle, like the sleeve of a wet jacket. Then a rattle of hasty steps across the slick scree, hurrying in the direction of the castle.

Helias reached the path. He looked around, or at least as far as the fog allowed.

An orange laser beam cut through the murk ahead of him, illuminating a package on the path, a few paces away. The light went out.

Helias picked up the waterproof package. He opened it. He found a semi-transparent plastic tube, sealed with a soft stopper. There was a strip inside, the size of a small bandaid, made of a strange pliable material. There was also something written on a card, which he could barely read in that light. He put everything in his pocket and made his way to the castle.

In his room, Helias took the strip and placed it below his ear, behind the angle of his jaw, as the instructions said. He went to the mirror. It adhered perfectly and his hair made it almost invisible. He was supposed to say his name backwards, in an undertone, to activate the communication.

"Saileh."

He heard a buzz.

"Yes?"

"I'm waiting for instructions."

"We are communicating on a frequency that cannot be intercepted. In addition, the signal is coded and only our two devices can decode it. So you can relax, but speak softly."

It didn't seem to be Mattheus's voice.

"Who are you?"

"I am the professor."

"I don't know you."

"You will know me soon. Do not be mistrustful. Are the names of Mattheus and Kathia enough for you?"

"I don't know them."

"And the word, 'Thaýma', what does it mean to you?"

"Nothing."

"Okay. I will call you back in a moment."

The buzz again. Helias turned the device off by placing a finger on it.

The professor. And who might he be? Why didn't he say his name? What was his relationship with Kathia and Mattheus? If he was a friend, why did he answer and not they? If the professor was an enemy, then he himself had been found out. They had seen him on the north shore. And they were setting a trap for him. But then why hadn't they gotten him immediately down on the shore, where they could have easily taken what he had found, if they knew about it and if that was what they were after. No, he was letting his imagination get ahead of him. He just had to wait.

He was undecided whether to hide the diskette immediately, or have a look at it first. It wasn't compatible with his portable. Maybe there was a message he should read right away. He locked the door. He logged off from the network and loaded the diskette in the computer.

It was almost entirely filled with a password-protected compressed file. And then there was a message.

"Password for the message: 'The name of the brunette you liked, more than just a little, at college.'"

Well, she's the only one who could know that, since she had been 'stalking' him then. Was she jealous? The thought made Helias smile.

He typed in the password. The message appeared.

"If this diskette is in your possession, it is because you have to take over from me. It is no longer safer in my hands. Do not worry about what it contains, which would be incomprehensible for you in any case. Guard it well, however, it is extremely precious. Hide it in a safe place, where only you can find it. Never think about where you hid it, especially when other people are present. Neither I nor anyone else must know. Not even the professor, who will contact you to give you explanations. In case I have not been able to give them to you in time. Yours."

"Saileh."

"Yes?"

"It's me. Sorry to have been suspicious."

"No, not at all. You were right to be. I appreciated it."

"You need to talk to me?"

"Yes. We need to brief you on the entire business. You're not obliged, but it is advisable."

"I ask for nothing better."

"Good. Are you with us, then?"

Helias remained silent for a moment.

"I'll wait to hear what you have to say."

"Fine."

"How is Kathia?"

"Fine. Well enough. But you cannot see her. For the moment."

"Why?"

"Come to me and you will know."

"We have to meet in person?"

"Yes, it's better."

"But isn't it risky?"

"Not particularly."

"Okay, then. Tell me what I have to do."

"My office is on the west side, third floor, almost diametrically opposite your room. Look for me on the map, Professor Borodine. Don't cross the courtyard, but reach my office through the internal corridors. Don't rush. Take all the time you need to make sure you're not watched, especially as you enter my office. Change route several times, if necessary, or go somewhere else if you suspect you're being watched. Leave the device on, but speak only if necessary and where no one can see you or hear you. The device also contains a signaling unit and I can track your movements on the map of the building. Is everything clear?"

"Yes. When?"

"Immediately, if you like."

"Let's say in ten minutes or so. I'll call again before starting out."

"Good, see you soon."

"See you soon."

The professor passed his finger behind his ear.

"Are we going to tell him everything?"

“Well, not exactly everything. Obviously.” answered Mattheus from his position at the window, as he gazed at the reflection of the yellow and orange lamps on the rain-roiled surface of a puddle in the courtyard.

Chapter 6
Seated Behind His Enormous Desk

Seated behind his enormous desk, the professor had a face he didn't like. He especially didn't like those two flaps of skin that hung down from his cheeks and below his chin. They made him look like a bulldog every time he moved his head, jowls flaccidly jiggling.

He wasn't actually all that ugly, really. If it weren't for the bags under his eyes and those dog-like dewlaps, he wouldn't have looked too bad.

Maybe he drank. Who knows if the Thaymites drank? He certainly didn't belong to a 'genetically enhanced' breed. Assuming that they used biogenetics for that on Thaýma. After hearing Kathia, one would have said they didn't. Though one look at her would lead to a different conclusion. Eh, who knows?

"No problem in coming here, young man?"

Aargh. The situation was getting worse. He hated being called 'young man'. He was twenty-seven years old, almost twenty-eight. He tried to take it as a compliment. Not very successfully.

"I noticed there was just one small detour, and pause, along the way...."

Worse and worse! He seemed to be doing everything he could to make himself unpleasant.

"Nothing in particular. You know how it is.... Maybe the wet weather...."

The professor didn't seem to understand.

"You know.... The rain.... Probably, certainly, in fact, tension too.... Emotional...."

The professor studied him closely, tugging at his chin with thumb and forefinger, squinting. Then he said "Ah!".

But what did he care about his bladder? Helias was verging on furious. What was this? He was supposed to report on every time he went to the toilet? Did they want the details too? A video to see whether it was light or dark? Straight or helical? He already wasn't any too pleased about that whole business of Kathia's 'surveillance' of him just about everywhere. But with Kathia it was different. He didn't mind not having secrets from her. But this, this...with those wattles of his... how dare he?

M. Villata, *The Dark Arrow of Time*, Science and Fiction,
https://doi.org/10.1007/978-3-319-67486-5_6

"Let's get back to us, young man."

Grrr.

"Just a moment. First tell me about Kathia."

"As I've already told you, she's fine. But you cannot see her. Above all, because she could involuntarily 'read' your secret. You know what I'm talking about, don't you? You found the object and you hid it?"

"I'd rather change subject."

"Ah, I understand. No, you don't need to be afraid of me. I'm not like the people you know from my planet. I don't read other people's minds. I have enough trouble with my own...."

And he laughed. And when he laughed his jowls bounced horribly up and down.

"Go ahead and relax. There's no danger...."

"How do I know I can trust you?"

"All right. Let's go over there. First, though, please turn with your back to the bookcase for a moment."

The professor's office was almost entirely furnished and fitted out in dark wood. Most of the left wall was occupied by an enormous bookcase that went from floor to ceiling. It appeared to contain any and all kinds of document, the widest range of forms and supports, analog and digital, disks, diskettes and cards, hard and soft, cassette tapes, large and small, printed books in many languages. And all from every age and place.

"Admiring my collection? One of these days I'll show you the rarest pieces. The tapes, obviously, are unusable. I keep them for the covers...."

Okay, okay. Let's just cut to the chase. Helias turned his back. He heard the professor fussing around the bookcase. When he was allowed to turn around, one bay of the bookcase had been transformed into a door opening onto a dark room.

"Come. Light. Screen. Door."

The light and the screen went on, the 'door' closed behind him.

"Mattheus?"

After a few seconds, the image of Mattheus's face appeared.

"Yes?"

"Mattheus, could you please confirm to this young man that I don't 'feel'?"

"Hello, Helias. Yes, that's right. You can trust him, in all senses."

"Where's Kathia? Is she there with you?"

"No, she's not here. As soon as I see her, I'll give her your regards. Goodbye."

"Wait...."

But the communication had already been broken off.

"So, young man. Sit down. Let's get back to business."

Helias sat in one of the armchairs arranged in a half circle around the screen, now off.

"Professor, could you do me a favor?"

"Tell me. If I can...."

"Don't call me 'young man' anymore."

The professor stared at him, astonished, with his eyes bulging out of his head in an utterly idiotic expression.

"What? Why?"

"Because I find it very irritating."

The professor continued to goggle at him.

"Are you nervous, young man? No, what's your name again? Kadler? Are you nervous, Dr. Kadler?"

"Nervous? Me nervous? Of course I'm nervous! Why shouldn't I be nervous? Is there any reason at all that I shouldn't be nervous?"

Helias rose from the chair and began to pace back and forth, nervously.

"Give me a reason I shouldn't be nervous. Give me just one damn reason I shouldn't be nervous."

Throughout this rant, he had been stabbing his finger at the professor who, in the meantime, not knowing what to do, had sat down, embarrassed and flustered.

"A few days ago I said goodbye to everybody and everything. Just like that! Who gives a damn anyway! I get tossed onto this fucking planet. I don't even know how. I don't even know if I've still got all my cells in the right place. Maybe one fine day I'll realize some of them are missing, floating around in space. The minute I arrive I get practically kidnapped, oh, all very politely…but not really all that politely. Maybe I needed a little time to get used to things here. Who knows? I certainly don't, I didn't even have time to ask myself before I get caught up in who knows what fiendishly complicated plot, and discover I've got a mission or something like that, what it is I don't know, I don't know a thing. I find myself having to be a secret agent, without knowing for who or for what noble cause. Me, who never even did so much as sneak jam from the cupboard as a kid. I get three or four good frights, from people who materialize from behind the walls, or creep around in the dark, dragging their feet in the fog. And you want to know if I'm nervous? Damn right I'm nervous. Very nervous. Nervous, pissed off and tired. Good lord."

Helias sat back down, head in hands. The professor was about to say something, but stopped halfway, his mouth ajar.

"A beautiful woman, from another planet, says she loves me. She says she also knows how many hairs I've got on my head. And to know what I'm thinking. Even my most intimate thoughts. A woman like that can steal your soul, don't you think? Or frighten you. Or both. I don't know. All I know is that she's always in my thoughts. Like an obsession. A wonderful obsession. Or an anchor, a haven. Because she's all I've got, in this whole crazy story. And now she's gone. I can't even so much as see her."

Helias fell silent for a moment, eyes downcast.

"Sorry. I didn't mean to tell you these things. Maybe I just needed to get it off my chest…."

"You're tired. Have you eaten?"

"Not much, I wasn't hungry."

"Would you like something? An ice cream, perhaps?"

"I don't know, I might…."

"Wait for me here a moment. I'll be right back."

The professor went out for a few minutes.

"Here's the ice cream. I always keep a stock in the freezer."

Helias looked at the ice cream. He took it. And he devoured it.

"Thanks. Could you tell me where the lavatory is, please?"

When he returned, the professor was lost in thought.

"Mattheus." he said, addressing the active screen.

"Yes?"

"Mattheus, could you please check whether Kathia is available? Tell her someone would like to see her."

"She's in her office. I'll have her call you back shortly."

"I'm going to the other room for a moment, young… Dr. Kadler. I have a few things I have to do…."

And for the first time he looked him in the eyes, and smiled.

Helias looked at him in turn. Actually, those bags under the eyes weren't that ugly, really. The wattles, no, those he really couldn't take.

"Thanks." murmured the young man.

Kathia's face was drawn. She was trying to smile, but couldn't hide her fatigue and worry.

"Hello, Helias."

"Kathia, how are you?"

"Fine."

"Are you sure?"

"Yes, I'm fine."

She seemed embarrassed. And she kept glancing off to the side, as if there were someone beyond her screen.

"Is there someone with you?"

She nodded assent.

"A little further over there."

"Can you talk?"

She essayed a pallid smile and dropped her eyes.

"More or less."

"What's going on?"

"The professor will explain it to you. We can't talk much. The line is secure, they can't listen in on it, but if we use it for long they might be able to discover it and trace our location."

"Who?"

"We don't know. Or rather, we're not sure."

"Are you in danger?"

"N…no. Maybe…."

She had lowered her eyes. She was lying.

"Is there anything I can do?"

"You're already doing a lot…."

"For you, I mean."

"Yes, Helias, you can do a lot for me…."

"What?"

"Things, unfortunately, have spun out of control…."

"What is it that I can do?"

"You can look at me, Helias. Look me in the eyes. For these few seconds that remain to us."

Kathia had brought her hand close to the screen. Helias did so too. Their fingers seemed almost to touch.

Helias looked her intensely in the eyes.

"Be seeing you, Helias."

"When will I be able to see you again?"

"Soon, I hope. But I still have to find a way to do it."

"What do you mean?"

"We've run out of time. Look at me again."

Kathia smiled. But her eyes were sad. Then the image vanished.

The professor reappeared after a few minutes.

"How's it going now?"

"A bit less nervous. But upset."

"Perhaps what you need is a good meal, and some rest."

"Is she in danger? How much? In what way?"

"We don't know. In any case, she's safe where she is. At least for the moment. She might have to leave. They may have found her, but we don't know what their intentions are. We don't even know exactly who they are. We're trying to understand. Sooner or later they'll have to come out in the open. We're hoping they'll make a false move. For now, we only have the 'sensation' that Kathia has been found out. And so we have to be extremely cautious. We had to 'isolate' her, or rather, get her entirely out of the picture, as if she had never existed. There's no longer any trace of her, not even in the archives. If our sensation is wrong, so much the better. When we're sure, she could even come back. Conversely, if our suspicion is grounded, they're looking for her and they'll have found she's been 'erased'. They could ask for a 'back-up' of the archives, but that would tip their hand, so they probably won't do that. We don't know whether they are informed of our station in the mountain, nor whether they know the people Kathia had dealings with, like you, Mattheus, me and the others. All we can do is reduce the interactions between us to the bare minimum, and not put ourselves in situations where someone could easily read our thoughts."

"How would they have been able to find her out?"

"Are you sure you don't want to rest a little, young… Dr. Kadler?"

"Perhaps you're right. When can we continue?"

"Tomorrow, I'd say. You need a good supper and a good night's sleep."

"All right. I thank you."

"I'll see you as far as the corridor. Remember, don't let yourself be approached by anyone, if you can avoid it, not even in the dining hall. If you can't avoid it, think as little as possible, or even better, think of 'innocent' things, confusedly."

They returned to the office and the professor turned on a monitor.

"There's no one in the corridor. Wait, I'll go out first."

The professor, unexpectedly, offered his hand. Helias shook it.

"So, see you tomorrow, young.... Sorry...."

"It doesn't matter. Go ahead and call me whatever comes naturally."

"Turn on the device. I'll 'follow' you back to your room. When you go out again, leave it behind, it's more prudent."

The professor, after one last look at the monitor, went out in the corridor.

"All clear. See you tomorrow."

Helias went straight back to his room. No detours, even when passing the lavatories.

Helias had returned to his room. Shortly thereafter, he had gone out for supper, steering clear of other people as much as possible. He had had a hard time falling asleep, and when he did he slept uneasily at first. Then he had slept heavily until eight thirty, and woke feeling decidedly refreshed, ready for new battles. After breakfast he had contacted the professor and gone to his office, with the usual precautions. Here the professor had explained why they suspected that Kathia had been found out.

The previous morning Kathia had gone to the optometrist's office for her appointment. There was a younger doctor there, not the same one she had seen at her earlier eye examination. Perhaps the latter had taken the two days off for the 'jaunt' to the capital and asked his assistant to stand in for him. Kathia had explained her problem and the young optometrist, without saying much, disappeared into the adjacent office for a few moments. Returning, he was suddenly talkative, chattering away about everything and nothing, as if to distract attention. He was looking at Kathia's eye in the screen, talking nonstop, when the girl had felt, in the midst of that smokescreen of words, a thought, something along the lines of 'Bingo! Finally!'. Kathia had had the distinct sensation that she had been found out. The examination was over and the young doctor was by now in full verbal flood, but they both watched each other guardedly, trying to conceal their suspicion. "Come back in two days, the new lenses will be ready." the doctor had said as he handed her the card with her appointment. "And the other ones, what was wrong with them?" Kathia had asked, but the doctor had purposely framed his answer in incomprehensible technical jargon.

"Will you be here in two days?"

"Not necessarily."

In practice, he continued not to answer.

"Your name, please?"

The young man was still looking at her guardedly, without responding. Apparently his thoughts were elsewhere. At a certain point Kathia 'felt' that he was ogling her, as if he had stripped her bare with his thoughts and then started to touch her. She knew she mustn't react: it was a trap designed to unmask her once and for all.

"For reference. In case you're not here."

"My name is on the door, next to my colleague's."

"All right. Goodbye."

As she went out, Kathia had taken note of the two names on the door, calling Mattheus without delay. Mattheus had immediately checked the archives: neither of the two doctors matched the description. Then they had broken off the call. Kathia had gone to her quarters, to gather up everything that might be compromising. In the meantime, Mattheus had 'erased' her from the archives, and checked that no one had asked for information about her. And then Kathia had vanished inside the mountain. And Mattheus had gone to her room, to remove all trace of her presence.

The professor had then asked Helias to tell him what he already knew, and Helias had given him a brief summary.

"So you don't know what the object you're safeguarding contains?"

"No."

"Are you interested in knowing?"

"I'm interested in knowing everything. From A to Z."

"Very good."

They were in the 'secret' room. The professor checked the time.

"It's a long story and it's almost time for lunch. Let's wait until later. You think better on a full stomach."

Looking at the professor's prominent paunch as he rose, Helias mused that he must do a lot of thinking.

"So, down to business, young man."

The professor had assumed a markedly academic air, most likely induced by ongoing digestion. The term 'young man' must have been part of his lecture-room manner, and evidently an inseparable part.

He had had a very abundant meal brought in, sharing it, rather sparingly, with Helias. Afterwards he appeared to have blissfully put all fleshly desire to rest.

"Right. You remember your Lorentz transformations, young man?"

"Vaguely."

The professor took a pen and a tablet and wrote something.

"Screen."

The writing on the tablet appeared on the screen.

"$x'^{\mu} = \Lambda^{\mu}_{\nu} x^{\nu}$"

"Yes, I remember. That's the covariant formulation."

"Remember this too? Which derives from the quadratic form's linearity and invariance requirements."

"$\det \Lambda^2 = 1, (\Lambda^0_0)^2 \geq 1$"

"More or less."

"Now for the important part. If we disregard the improper transformations, or those with negative determinant, we are left with the proper subgroup, where '$\det \Lambda = +1$'. The latter includes the orthochronous transformations '$\Lambda^0_0 \geq +1$' and the antichronous transformations '$\Lambda^0_0 \leq -1$', which do not preserve the sign of the time component. Like the improper transformations, the antichronous transformations are also usually considered to have no physical meaning, or in other words, as not applying to any phenomenon of the physical world as we know it."

"Unless we consider the existence of antimatter."

Clearly startled, the professor swiveled toward Helias, a pleased look on his face.

"Exactly, my boy, exactly. But how...?"

"I don't know, maybe I heard people talking about it in the past. Or maybe I just connected it with what Kathia was saying...."

"Ah. They had told me you were an intuitive type, son."

'Son'? 'My boy'? What was happening to the professor? Not just vocabulary, also the way he looked at Helias had changed. Now he was looking at him with a sort of scientific curiosity, pleasantly surprised.

"Go on...."

"No, no. You go on, professor, please."

"Right. Take the quadratic form '$d\tau^2 = dx^\mu dx_\mu = \eta_{\mu\nu} dx^\mu dx^\nu = dt^2 - d\mathbf{x}^2$' and its canonical conjugate '$m_0^2 = p^\mu p_\mu = \eta_{\mu\nu} p^\mu p^\nu = E^2 - d\mathbf{p}^2$'. They are invariant quantities—scalars—of the theory. Now let's take something traveling at speed '$v = c = 1$', where c is the speed of light in vacuum (equal to one in our units, where space and time, as well as mass, energy and momentum, have the same dimension), invariant in any inertial frame. These particles thus have zero proper time and rest mass: '$d\tau^2 = m_0^2 = 0$'. In addition, they cannot be considered at rest in any reference frame, as their speed is always c."

"Okay. This is something I've never understood. How can 'something' not have its own reference frame?"

"Yes. In reality, it's a sort of dialectical contradiction, but it's only apparent. The contradiction is resolved by the plusvariant formulism, which you don't know, where the reference frames are 'replaced' by reference 'states', which take all the intrinsic variables into account, for the real as well as the imaginary parts. The reference frames are simply trivial approximations of a simplified 'metric' for these states. Obviously, I can't go into the details of a formalism you don't know. We'll talk about it again in the future, if you like. Consider, in any case, that a hypothetical photon at rest would have zero proper time, as well as zero mass, as we've just seen. It would be a reference frame in which everything, space, time and energy, is reduced to zero, becomes unmeasurable: and what kind of reference frame would that be? But let's continue with the formalism you're familiar with."

Helias had coiled himself up in the armchair, rubbing and tugging his nose.

"For the moment, we'll ignore the singular case of photons, the case where '$d\tau^2 = m_0^2 = 0$'. We'll consider the other particles, the 'normal' ones, that can only travel at subluminal speeds, less than c. They thus have '$d\tau^2 > 0$' and '$m_0^2 > 0$'. In particular, we're used to dealing with positive proper times and masses. But nothing prevents there from being particles with '$d\tau < 0$' and '$m_0 < 0$'."

"Antimatter, in other words?"

"Precisely, except that in Lorentz's and Einstein's day, antimatter was not known to exist, and so there was no awareness of this kind of particle and the associated antichronous transformations. Later, when antimatter was discovered, it was more convenient—with the mathematical formalism used in those days—to consider the antiparticles as strange particles, opposite to the known ones, but still 'traveling', together with us, forward in time. Instead of considering them as particles that are perfectly normal, but travel in the opposite time direction. It was only a question of convenience in describing physical events mathematically. Or even an involuntary, primitive refusal to consider that two arrows of time could exist? A sort of unconscious repugnance? A bit like when astronomers didn't want to acknowledge that the Sun was at the center of the solar system, which made describing the motion of the planets so simple and direct, but preferred to 'invent' curious and complicated epicyclic motions, all so as not to change reference frame, all so as not to admit that the reference frame they found themselves on was not what the preferential one. A sort of intellectual laziness? Or a recondite—but not all that recondite—ancestral anthropocentric wishfulness? Was it the same for there being two arrows of time? Why was it so hard to admit that there was another one, that wasn't ours? And that everything had to be 'viewed' and 'measured' from what was in the middle, from the standpoint, or system, or reference state from which everything is most symmetrical? The Sun, in one case. The 'photon' in the other, because it's in the middle, between the positive and negative masses, and doesn't need space and time in order to exist."

The professor was heating up, his eyes flashing with proselytizing fire. By now, Helias was propping his chin up on his knees.

"The plusvariant formalism and the Copernican revolution thus have the same meaning in the epistemology of physics. Just as the solar system can be simply and symmetrically described if it is 'seen' from its center of mass—the Sun, approximately—which can be considered stationary in space relative to the system, so the physical world can be viewed and described more clearly and symmetrically from the standpoint of the photon, which, relative to us, is stationary in time. It's a sort of bank standing high and dry between two canals that flow in opposite directions. A bank you have to cross if you want to reverse your direction in time."

The professor lapsed into silence, and stood contemplating the tips of his shoes. Helias, fascinated, hardly dared draw breath.

"Are you familiar with the CPT theorem in relativistic quantum mechanics, young… Dr. Kadler?"

"Y…yes…. I believe I've understood…."

"Excellent. Tell me…."

"If I recall correctly, a CPT transformation…. No, let's put it this way: the laws of physics are invariant for CPT transformations. In other words, if we have a system or an event and we apply a charge conjugation to it, and also a parity operator and a time reversal, we obtain an event described by the same physical laws."

"Yes, you've expressed it poorly, but the concept is more or less that. And so?"

"And so I'd say that CPT symmetry is closely related to what you've said so far. If I'm not mistaken, it would mean that the description of an event in which a particle goes backward in time is equivalent to one where an antiparticle goes forward in time. But I don't remember what parity has to do with it. And what's the connection between charge conjugation and antichronous transformations?"

"Well, without going into the details, the CPT operation corresponds to the antichronous transformation called total inversion '$-\mathbf{1} \equiv \mathrm{diag}(-1, -1, -1, -1)$' combined with the reinterpretation principle. At this point, then, we should be convinced that the two descriptions, the particle going backward in time and the antiparticle going forward in time, are entirely equivalent, at least as far as special relativity and quantum physics are concerned. But what happens if, instead of considering single particles, we consider a macroscopic system of particles, one that's also subject to other laws, like those of thermodynamics?"

By now, Helias was gnawing on his cuticles. Something he had never done in his whole life.

"Right, so if I've understood correctly, the two situations, the antiparticle going forward in time, and the particle heading toward our past, are indistinguishable from the standpoint of special relativity and quantum mechanics, precisely because at that microscopic level it's not possible to identify a 'preferential' time direction, an arrow of time, as on the contrary is the case for macrosystems that have precise thermodynamic evolution, the only one possible, marked by increasing entropy."

"Exactly. That's the concept, though you've put it a bit picturesquely."

"There's another thing I've not quite grasped: why total inversion? Why invert all three spatial axes? For a parity transformation, which in practice provides a mirror image of the phenomenon, isn't it enough to invert just one axis?"

"To a certain extent you're right. You can invert all three or only one, but not two: the transformation would be improper, it would have a negative determinant. In the world of elementary particles, it doesn't make any difference whether you have one or three: there's nothing that distinguishes between 'up' and 'down'. But if we take a system with non-negligible mass and we 'switch on' a gravitational field, then it makes sense to invert the other two axes as well."

"So what you're telling me is that if there were a macroscopic system going backwards in time relative to us, we would observe it not only with a reversed time evolution—as is to be expected, obviously—but also as a mirror image and upside down. Let's take an example. Say we have a pot, a plain old-fashioned pot, the sort of thing you'd put flowers in, made of earthenware or glass, we put it on a table, on the left edge of the table, and we somehow send the whole setup back in time. What we observe then is an overturned table, with the pot 'glued' to it, upside down, on the right edge of the table which is now on our left. Or no, actually, seen from behind. Unless we walk around it. Is that right?"

"Yes. Except for the fact that if we, now, send it back in time, we would stop seeing it, whereas we would have seen it in our past. Go on."

"Of course, you're right. So let's say that our table-pot system (and while we're at it, we'll also add a floor) comes from our future. Suppose that at a certain point, still in our future, say, in a few moments, an accident of some kind makes the pot fall off the table, and it shatters on the floor. And so, in a few moments, we will see all those broken bits collect together on the floor above the table and reassemble into a perfectly sound pot that would calmly rise off the floor and sit on, or rather under, the tabletop. Is that right?"

"Very good. You've just given a description of the principle of general CPT invariance, called CST invariance, whereby all the laws of physics, and not just those of quantum mechanics, are invariant for joint CST, transformations where T is the time reversal, and S the space inversion, which, together, correspond to the Lorentz total inversion '$-\mathbf{1}$', and where C is no longer only the charge conjugation, but the scalar quantities, like rest mass and proper time, also change sign. Consequently, our anti-pot, repelled by the gravitational field, sits happily below its anti-table, perfectly comfortable. And up to here the description 'antimatter going forward in time' would still be valid and indistinguishable from matter going backward in time. The same is true if our anti-pot were unbreakable and its impact against the floor were perfectly elastic: we would have an anti-pot that, after infinite equal rebounds, would come to rest exactly on the table, transferring momentum and kinetic energy to the elastic body that made it fall. Strange, but mechanically possible. But if the impact is not elastic, and the pot breaks or at least is distorted, or in other words if energy degrades and entropy increases, changing sign for '$d\tau$' would indeed save the laws of thermodynamics and everything would be formally okay, the formulation of the law is not changed, but can we still believe it? Do we still believe that we are dealing with antimatter that travels with us in time? Would an anti-floor really have the power to challenge all the laws of probability, giving each of the anti-pot's fragments the right momentum to go reassemble in the exact position where the anti-molecules would stick to each other, and finally give a good kick to the whole pot so that it comes to rest on the anti-table? No living anti-being could do that, let alone mere chance. Another arrow of time is the only, obvious solution."

The professor was visibly tired from the discussion. Helias even more so.

"An ice cream, my boy?"

Helias ate half of his ice cream in silence. The professor had already finished his. He had devoured it, practically attacking it.

"When did you discover all this? I mean, when were you able to 'produce' enough antimatter to allow you to observe its thermodynamic behavior?"

"A long time ago. And thanks above all to the plusvariant formalism, which enables us to find the most direct way of exchanging energy with the photon state. In addition, we already knew that it was a question of reversing the time arrow of something that already existed, and not of producing something new, and this was a major advantage."

The professor remained in silence for a while, bowl in hand. He seemed to be weighing whether to have another helping of ice cream or not. His conscience quickly got the better of him, and he put the bowl back on the table, with an air of regret. He stood musing for a little longer, looking down. Then he raised his face to Helias, eyes sparkling in amusement.

"You gave a very nice example, with the breaking pot. Now I'm going to give you one that's even more challenging. Let's say we have an anti-man. Rather, it's you and me, in the same room. Each of us is going toward his own future, but in opposite directions. Each of us is aging, but sees the other getting younger, head where the other's feet are, and heart on the right hand side. The room, let's say, is on my side, together with the air it contains. You have your armchair and the air you need to breathe. The whole business, obviously, is surrounded by magnetic fields that isolate the two systems, otherwise they'd wipe each other out. You've arrived a short time ago, where you came from and how you got here doesn't matter. In reality, when you appeared you were going away. I know now that you'll be leaving in a few minutes, because I saw you do it a few minutes ago. But I don't yet know anything about our meeting, or how long it will last. You, however, know everything, because it belongs to your past. As my time passes, I will come to know about the meeting, while you will lose that knowledge. We look at each other, but I don't know if we can communicate verbally. Let's suppose we can: the magnetic fields can transfer the sound waves. At a certain point, my eardrums begin to vibrate: they're the last things you said to me, the words are backwards, incomprehensible. Perhaps they reach me even before you pronounced them, it depends on how far I am from the magnetic field, beyond which the sound waves travel backwards, until they make your vocal cords vibrate. I don't know how long you've been talking to me, but I record it all, so that I can look at it and listen to it at my convenience, backwards. You talk for a long time. You've probably repeated the same things, because otherwise I would have missed the last parts, before starting to record. At a certain point you stop talking, that's the moment when you had started. I stop recording, very noticeably, so that in this moment you will know I recorded you, or rather, that I will record you: it's the signal for you to begin talking. You do the same thing, you turn on the camera, or rather, you've just turned it off: and that's the signal for me, it means that I will speak to you, rather, I've spoken to you. To make sure, you show me a sheet of paper where you've written, 'Now you talk, and then repeat what you said'. Then I begin to talk, I introduce myself and I tell you a few things, twice. A bit before I finish, you turn off/on the camera, so that I know I am about to finish, and in fact that's the way it is, because you, like me earlier, have turned on the camera after you heard me finish talking. And so, after waving to me, you leave relative to my timeframe, whereas for your own, you're just arriving. Right. I'll let you think about this for a while. When I come back, I'd like to hear your comments about this business."

The professor, taking the two bowls, left the room. Helias knew what he was going to do. And he took advantage of the time to go to the lavatory.

The professor was still licking his lips when he returned. Helias had curled up again in the armchair.

"First, I'd say that we would have communicated more efficiently in writing, with messages prepared and left at appropriate times. This, though, is only a technical detail that doesn't change the substance of what you described to me. But before I explain my doubts to you, tell me if things like this actually happen."

"No. At least to date, and as far as I know. As you've been informed, we've forbidden ourselves from interfering with the past. So if this prohibition continues in force, we cannot receive information from the future. It was just a little intellectual exercise, invented on the spur of the moment. There are people who study these things much more seriously and in greater depth. But we can learn something even from this little example. Let's hear what you have to say about it."

"Well, I'm certainly quite perplexed. All of this clearly runs counter to every notion of causality and free will. Not that this frightens me, mind you, but for most scholars on Earth this would all be repugnant, to say the least. Take, for example, the message, 'Now you talk', and so forth, that I show you at a certain point. But why do I show it to you? I already know that you've talked, or that you will talk, I've even recorded you. For me, it's already happened, by now it's part of an indelible past, I've also got the proof. On the one hand, what need do I have to urge you to do it if I already know that you'll do it anyway? On the other hand, if I don't do it, you might also not decide to talk, even if that seems impossible. In other words, you have no choice, you will definitely do it, regardless of what I do, and I can always decide not to show you the message. And this is just an isolated example. In actual fact, the whole thing seems to be like an unalterable script that both of us have to stick to, without being able to do otherwise: what's become of our free will? So it doesn't exist? Is everything already in the script and the only thing we can do is to act our part and speak our lines? So it would seem, since the two of us can't do anything except what the other one has already seen us do."

"Good. You've hit the crux of the problem. Except that you're not able to shake loose a 'single-time' outlook that sees the future as something that is yet to come and can still be changed. Let me explain that better. In our example, I, and I alone, have chosen to speak, I didn't stick to a script that was already written. You induced me to speak: it was your future choice to induce me to speak in your past. Both of us made our choices, and the result is what happened. We just have to liberate ourselves of the idea that causality has only one time direction. In other words, causes lie not only in the past, but in the future as well. What you chose in the future—to show me the message—influenced what happened in your past, it induced me to speak. It's still a matter of cause and effect, even though in one of the two time directions it is necessarily seen the other way round. It always happens when you're dealing with time travel, of any kind. Take the famous 'grandfather paradox', where a time traveler has to choose whether to kill his own grandfather when the latter is still a little boy: were he to do so, he would never have been born and he would not even have been able to kill his grandfather. The solution is not that time travel doesn't exist, but simply that the time traveler chose not to kill his grandfather: a future choice that affected his entire past, his very existence."

They both remained in silence, while Helias pondered all those new and amazing things, trying to assimilate it all. It was the professor who spoke first.

"There's another important physical problem hidden in our example, which you may not have noticed. We said that the two of us, you and I, speak to each other in some way. In other words, we assumed that the sound waves can somehow propagate from one system to the other. Let's not worry for the moment about the details of crossing the magnetic barrier, but focus on 'before' and 'after'. You speak, and the sound waves produced by your vocal cords propagate around you, moving away from you as your time moves on. Simultaneously, as we said, if I could visualize the phenomenon I would see those waves traveling backwards in your direction until they make your vocal cords vibrate. What happens when the signal 'passes' into my system? Well, it must necessarily obey the thermodynamics on my side, in accordance with my time arrow. And so it will propagate normally, forward in my time, and I will receive it a few instants after it passes across the barrier."

"Yes, I had started to think about that, but then I got caught up in the other problems. What happens with the light signals or, in general, for electromagnetic waves?"

"Excellent. And that opens a whole new chapter, and a crucial one. As you may recall, Maxwell's equations contemplate both retarded and advanced solutions for the propagation of electromagnetic waves. In other words, waves can be emitted both forward and backward in time, with positive energy and negative energy respectively. Accordingly, advanced radiation reaches us from the future, but we can't see it because its energy is opposite in sign. Not only, but for us it would be radiation that takes the reverse path: from the absorber to the source. Remember the Wheeler-Feynman absorber theory and Lewis's paradox? No? It doesn't matter. In other words, the photon, as we know from its '$d\tau = 0$', does not distinguish between past and future, it's arrived even before it's left, whatever the distance that we see it cover. It's just us, who belong to space-time, who are obliged to separate its presence into space covered and time taken. Here is where the constancy of the speed of light arises: it's space and time separating, with a dimensional ratio equal to c, equal for everyone. We detect the photon in our future because that's where it has positive energy in our time direction. By contrast, somebody going in the opposite direction will see what for us is the invisible advanced radiation, because for him it will be the radiation with positive energy, coming from his past. The photon makes no distinction between the two arrows of time, it's the middle ground, present for both, that acquires positive energy according to the direction of time. Just as our flowing in time gives us positive mass."

"So we see so-called antimatter through its advanced radiation, which to us appears as retarded, and vice versa."

"Exactly."

Some minutes of silence before Helias spoke again.

"Kathia talked to me about 'back-now', going back in time through the light-years that separate us from the Earth. I think I can fit that into what we've said so far. At the same time, though, I have the feeling that I'm missing something. Maybe I'm a bit tired…."

"When I'm tired I use the Feynman diagrams, which can be read in both ways: matter and antimatter that go forward in time, or matter only, going forward and back in time, as Feynman himself did. Do you remember them?"

"Yes. More or less."

"Now, though, don't imagine we're dealing with elementary particles, but with something macroscopic, like our craft full of passengers, for instance."

Craft?… That's what Kathia had called it too…. Where was Kathia now?… What was she doing?…

The professor picked up the tablet and drew some lines and symbols.

"This simple diagram represents the transmission of the craft from Earth to Alkenia, along this world line which is partly baryonic and partly photonic. Along this other world line, whose photonic part is shared with the first line, we have the transmission of an equal quantity of matter backwards in time. 'Back-now' simply consists of replacing part of the anti-plasma with a small passenger craft."

"Who, however, would have to be already traveling back in time…."

"That's right."

"And how do you do that?"

Helias was having a hard time getting the words out.

"Look at this. One of the simpler Feynman diagrams. Then we'll take a look at something a bit more complex."

The professor continued to draw diagrams and explain them in detail, with no interruption from Helias.

At the eighth Feynman diagram, Helias fell asleep, without even changing position. Which is why the professor didn't notice until later. Four diagrams later. When Helias started snoring.

Chapter 7
A Sort of Autumn Had Arrived

A sort of autumn had arrived. The leaves on the little trees and bushes were yellower and drier. Some had already fallen, and were blowing in the bitter wind. But the sun was warm that morning, and Helias was savoring its warmth on his forehead and cheeks, seated on 'his' rock. The first rains had passed, and the clouds covering the mountains had gone, leaving snow-capped peaks behind them. And then they were back, and the intermittent rain was mixed with sleet. On the fourth day, the weather cleared again, and this time the snowfields' edge had dropped halfway down the slopes, an unwavering line scribed across all the surrounding mountains. It was spectacular, and Helias found it all inebriating, the scenery, the sparkling air and the bright sun.

The fresh snow and the angled rays of an ever more southerly sun had changed the landscape almost beyond recognition. And Helias had changed too.

He had changed department. He had been transferred, with his consent, to Physical Sciences, where his meetings with the professor would not arouse suspicion. And so he had had to cut himself off from the last thing that remained from his past: his research. However, he had managed not to change his room. It was a way of keeping something firm and certain, a memory of those first days, of that single, unique day spent with Kathia.

He hadn't seen her again, swallowed by the mountain. Where he couldn't go, so as not to reveal his secret.

He—an earthling—had to be the one to keep the diskette. An earthling was above suspicion. The professor had had it from a colleague, when he was on Thaýma.

This colleague was working in advanced experiments, dangerous stuff, pushing the limits of the permissible. Until then, transmission—whether forward or backward in time—required a transmitting station and a receiving station. Obviously. The new experiments investigated the possibility of transmitting anywhere, even if there was no receiving station. As the professor had explained to him, all of this was incredibly dangerous.

M. Villata, *The Dark Arrow of Time*, Science and Fiction,
https://doi.org/10.1007/978-3-319-67486-5_7

As it was now, 'back-now' made it possible to travel backward in time, from one station to the other. Only a few routes were allowed and transmissions in any case had to be duly justified. And the whole thing was rigorously controlled and controllable. Even if 'clandestine' trips—unauthorized ones—were somehow to evade notice, the 'time pirate' could at most go back to the period of the first existing station, and no farther. But what would happen if transmission were possible without a receiving station? Total chaos. A time pirate would be able to go anywhere and to any time, simply with a series of 'rebounds'.

This is why the project run by the professor's colleague, who was called Nudeliev, was not only top secret, but really did verge on the prohibited, and one almost wondered how it had ever managed to get itself authorized and funded.

And in fact something had gone wrong. Evidently the possibility of 'free' transmissions was very tempting to someone who had found out about it. And indeed, the review board for scientific research had many members, including the professor, and despite the secrecy surrounding the project, someone could have leaked something about it.

However it came about, one fine morning the members of Professor Nudeliev's group discovered that their computer system had been hacked, even though it had been protected by the best security systems. Though the group's findings at this stage had to be regarded as preliminary, the software they had developed was virtually able to control the free transmission of small amounts of matter. It was quite likely that the software had been copied and was now in unscrupulous hands.

However, in addition to the computer security measures and other precautions—for instance, no member of the group had access to the entire system, but only to the part on which he was working—the project leader, unbeknown to the others, had thought it best to introduce further safeguards. There was a 'key', a decoder or compiler or something of the sort, that only the group leader and the professor knew about. The professor was not involved in the project, was not part of the group, and had no other role. He had simply been chosen by Professor Nudeliev, as an old friend, to be the 'keeper' of this secret, in case something happened to the project leader.

The software was already coded to begin with, in the sense that it was written in a dedicated language known only to the group. But computer experts would have been able to translate it without too much trouble. Without the compiler, however, the software was completely unusable.

Professor Nudeliev and Professor Borodine had agreed that once a day the latter would receive a signal meaning 'everything okay'. If he stopped receiving it, all he had to do was connect to his colleague's private computer and transfer the compiler to himself, leaving no trace.

And this is what was done when the system was hacked. Nudeliev, who at that point could take it for granted that his every move was watched, didn't have to lift a finger. Had he been 'searched' or, worse, forced to reveal the secret, the compiler was safe.

The project, predictably, was put on hold, pending clarification, and the leader was provided with a bodyguard.

After some time, on the eve of his departure for Alkenia, the professor met with Nudeliev and, by common accord, decided that the professor would take the compiler—the red diskette—to Alkenia.

Once on Alkenia, the professor felt it would be more prudent to get rid of the diskette, since his friendship with his colleague was well known, and any search for the diskette would have quickly led to him. And apart from that, it was by no means unlikely that the ex-group leader would be forced to reveal his name.

He talked about the diskette with two people he had long trusted, Kathia and Mattheus, and they agreed to take responsibility. But there was alarming news from Thaýma: Professor Nudeliev had been kidnapped and almost immediately released, but in the meantime there could be no doubt that they had thoroughly probed his mind.

At this point, Helias had asked how it was possible to keep all these secrets on Thaýma, since people could read each other's thoughts.

The professor had answered that they were well trained to hide secrets in their minds, and that in any case, there weren't really all that many mind-readers among them.

After his release, the scientist had said that his captors had given him something to make him sleep, and that he remembered nothing. Then he had notified the professor that he was almost certain they had not read his name, though he couldn't be sure they hadn't found out about Alkenia.

It had thus been necessary to take further precautions: the diskette must be entrusted to someone unknown who would never be suspected, an earthling. But one they could trust, one hundred percent. And so the individual in question was scrutinized at length, without his knowing. After which his application for a transfer to Alkenia was accepted.

In the meantime, as yet another precaution, the diskette had been on Earth with Kathia, and all newcomers to Alkenia were rigorously checked, especially those heading for the Kusmiri Center. Clearly, though, something had slipped through the net. Not only. At a certain point, the watchers became the watched. With the ruse of the allergenic contact lenses and the fake optometrist, the 'others' were able to check the new arrivals, to distinguish Thaymites from earthlings. The new lenses, in fact, irritated and inflamed the eyes of the Thaymites, who already had their own special mimetic lenses. And so Kathia, just arrived as an archivist, was found out.

Helias, during one of his conversations with the professor, had objected, "Why go to all this trouble to prevent something that probably never happened? I mean, history is what it is, we have no knowledge of people who went back to the past to corrupt it. Nobody ever went, say, to kill Napoleon Bonaparte's grandfather to prevent him from being born. Or even if they did, they clearly weren't successful, since Napoleon existed. Nobody ever came from the future to conquer and enslave a primitive humanity, thanks to his advanced knowledge and technology. If it never happened, what is it that we're supposed to prevent now? Or, conversely, if something happened, something we don't know about, how could we prevent it now? In other words, hasn't everything already happened, or not happened?"

"Certainly, certainly. You can't manage to shake off the common single-time outlook, and I sympathize with you. And yet it's simple, what happened or didn't happen in the past clearly also depends on us here and now. The fact of knowing, in part, about the events doesn't change anything. They also depend on us anyway. Remember the example of the anti-man who at a certain point causes something he's already seen happen? His past was affected by a choice he makes in the future. For us it's the same, we're here fighting so that everything happens in the way it's already happened. There's nothing wrong with that, I'd say. Furthermore, don't forget that—and this should matter to you even more—we still don't know what's going to happen tomorrow. Tomorrow morning we could wake up free and happy as we are now, or enslaved and oppressed by a future power. Or we might not wake up at all. What do you say? Do we want to risk, and do nothing to prevent it? Or is it worth fighting, now and in the future, so that that doesn't happen tomorrow?"

"Of course. You're right. The answer is clear. Now I understand better. After all, even if it's our past, our history, part of which we know about and take for granted, as something that's already taken place and is now frozen and unchangeable, it still depends on us, now and in the future. In a certain sense it's also part of our future, even though it's our past at the same time. You're right, all you need to do is break away from a single-time outlook. Conversely, it's a bit as if we also belonged to the past and can thus influence it. It hardly matters if we already know how it will turn out. What I still don't get, though, is why you Thaymites, with all your wisdom, approved such a dangerous project, whose consequences—which were pretty obvious and predictable—you're now having to fight."

"It's the old dilemma between Knowledge with a capital K, and the dangerous consequences it brings. The thirst for knowledge almost always wins, not least because optimism is part of human nature. If it weren't, the human race would probably have died out pretty soon. And it's also true that if something becomes doable, it's partly because we're mature enough to do it. The vital thing is that it not end up in the wrong hands. And here, in this case, is where we come in."

All in all, Helias was a little disappointed. It seemed he had been chosen for reasons of reliability. Nothing more. What he was being asked to do was to guard something precious. Hardly more than a watchdog or a performing monkey. He was none too pleased. Not least because, from what he had been told before, he had got the idea that his 'mission' consisted of something more active. They'd spoken of his intuition, spoken of it as something valuable and essential. But what good would it be in the task he'd in fact been given? Wouldn't any earthling have done just as well? Any earthling with a minimum of smarts? But maybe that's exactly what had happened: he was nothing more than the first dependable earthling with a minimum of smarts to come along.

A bit timidly, he had tried to sound the professor out.

"You see, my boy, you've got a great quality: that of asking yourself the right questions. That is the key to intuition. Not finding the answers, but identifying the questions. The answers come on their own, they're the consequences of the questions. But if the questions are wrong, what good are the answers? You think I'm beating around the bush, don't you? You're right. You asked me a question, the right question. And the answer will come on its own. I know what it is, but I can't reveal it to you, not now. In practice, you've asked me what your real role in this business is, because you've sensed that there's something else, haven't you? You've shown excellent intuition, as usual, by the way. You'll know the answer soon. In a few days. Be ready."

That 'in a few days' had become today. The professor had told him to come after lunch (obviously, since 'you think better on a full stomach'), hinting that the time had come, a critical juncture: something was about to happen.

But Helias had no intention of getting all upset in the meantime, of agonizing over whatever that something might be. And so he sat happily basking in the warm sun, enjoying the nip in the air, surrounded by the curtain of snow covering the mountaintops, entranced by the splintered reflections of the sun flashing in the lake's jagged wavelets.

It was lunchtime by now. Ten minutes to go. He couldn't tear himself away from that idyllic scene, that sensation of peace, peace as deep as it was difficult to recapture.

Reluctantly he rose and, after a mute farewell to Kathia, he started off toward the castle.

As he stuffed his hands in his pockets, his fingers hit the object the professor had given him, saying it was from Mattheus. It was shaped more or less like a large curved cigar, slightly rippled to fit in a closed fist. Once the safety was off, a slight squeeze between finger and thumb was enough to release a paralyzing ray that disrupted an attacker's nervous system. The harder you squeezed, the stronger the ray. Or so he had been told.

There was another person. Aside from him, Kathia, Mattheus and the professor, there was another person who knew about the red diskette. An important person, a member of the Thaýma Council. At the time the hacking incident was investigated, the project leader, before being questioned, had asked to speak with the counselor, his old and trusted friend. Together, they decided that it would be unwise to hand the compiler over to the authorities, given the skill shown by whoever had tried to steal the project's findings, and the likelihood of complicity in the attempt. They agreed that the very existence of the compiler must remain secret, and that it should be kept hidden by a few trusted individuals, acting privately, unofficially, at the margins of legality.

When Helias and the professor entered the secret room, the counselor was there, pacing back and forth beyond the semicircle of armchairs. He barely turned to glance at them, pacing still, hands clasped behind his back. Helias and the professor stood waiting. Then the man stopped, staring motionless at the floor for few moments. Finally, with a mechanical gesture, he raised a hand to brush his medium length white hair away from his face and gazed at the newcomers with dark, deep-set eyes.

"Hello." he intoned in a booming voice that seemed to command unstinting attention.

"Helias Kadler, I presume."

"Yes, that's me."

"Professor, have you already acquainted Dr. Kadler with the reason for this encounter?"

"No, I thought it best for you to do that directly."

"So, Dr. Kadler, you are now our guardian."

"So it would seem."

"I imagine you already know everything, more or less in detail."

"More or less, I suppose."

"Sit down, please."

"After you."

"Thanks, but if you don't mind I would prefer to remain standing."

Helias and the professor sat. There was a sheet of paper folded in half on the low table.

"There is one important thing that the professor has not yet told you. And you will understand why in a moment. It's something that was known only to the project leader, Professor Nudeliev, and a very few others who were directly involved. If the project was top secret, this other thing, like the compiler, was secret to the rest of the project. Even Professor Borodine and myself were informed by Professor Nudeliev only as a result of the hack. You have been told that—as we all believed—the software developed by the group was virtually able to control free transmission. This is true, but it's only part of the truth. In reality, the software had already reached the testing stage, and the first tests had been successfully conducted, unknown to us all. Though it was a question of transmitting small amounts of matter over small distances, it was nevertheless an almost definitive success for the project. These experiments were performed far from Thaýma and far from prying eyes, on a station orbiting around a planet."

The counselor stopped and looked Helias in the eye.

"So you're about to tell me that this meant that there is a second compiler, or rather, a copy of the compiler?"

"Exactly."

"Where is it now?"

"Disappeared. Together with the people who performed the experiments."

Helias had begun to blink rapidly, fingering the tip of his nose.

"And yet 'they' didn't get it. Otherwise what point would there be in my safeguarding it? When did this happen?"

"A few months after the first theft. The station was thought to be a safe place, and so the second compiler was still there."

Helias realized only then that he had no idea of the date of any of these events. His throat dry, he asked "Meaning? How long ago?".

The counselor was watching him closely, more and more closely. But the answer came from the professor.

"A bit more than three earth-years ago, my boy. Three years before you were transmitted to Alkenia. And now look at that sheet of paper. It's a copy of a message found at the station after the disappearances."

Helias stretched out a hand toward the table. When he touched the paper, his hand began to tremble.

The message was hand-written, probably in a hurry. The script was middling in size, slightly slanting.

"23 May ED—My son was here. He took it. My husband is wounded"

Evidently there had been no time to write more.

Helias had paled, and his hand would not stop shaking.

"It's your mother's handwriting, isn't it?"

Helias had to swallow several times before he was able to speak.

"W…what kind of a joke is this?" he stammered, looking bewilderedly from one man to the other.

"I'll get you a glass of water." said the professor, rising.

The glass of water arrived right away, but Helias, huddled with his head between his hands, didn't even see it. After a while he glanced up, took the glass and barely wet his lips.

"What kind of joke is this?" he repeated, staring into space.

The counselor, who had not taken his eyes off Helias for an instant and continued to watch him searchingly, answered "It's not a joke. This is really your mother's handwriting."

"And so you think that I took the compiler…. And maybe, while I was there, that I wounded my father…. Is that right?"

"We believe what your mother says. We have no reason not to."

Helias raised his head and looked at the counselor, the strain and suffering in his eyes frightening to behold.

"And why, supposedly, would I have done such a thing?"

"We are certain that you took it. How and why is something you have to explain to us. We'll leave you alone to think about it."

And, leaving, he took the sheet that had dropped to the tabletop.

Kathia, on the other side of the partition, suffered silently.

The counselor handed the sheet to the professor.

"Measure it, please."

The professor took the sheet with caution, placed it on the desk and held a sort of sensor over it. Then he looked at the computer.

"Full scale." he said.

It was all so absurd. Why was everything so completely absurd?

How could they even think such a thing? Didn't they know him well enough? Hadn't they read his thoughts enough, peered into practically every nook and cranny of his brain? Didn't Kathia know everything there was to know about him? Or didn't they trust her either? Where was she? She, and all that talk about reliability? That, more than anything else, was the absurd thing: first they say they know him better than he knew himself, and then they come out with all these groundless accusations.

But they had been categorical: "We are certain that you took it.".

Or was he the one who didn't remember? Had they removed it from his memory? Or maybe it was…. But of course! It must be like that! That was the only possible explanation.

The counselor and the professor still had not returned. By now, Helias had recovered from the state of prostration he had fallen into.

The strange thing now was that he realized he hadn't been especially shocked to hear that his parents had been involved, as he was himself, in this business of the compiler. Had been, or were involved? A shiver ran through Helias. The counselor had said they had disappeared. And if they were still alive? Perhaps prisoners? A sudden agitation threatened to overwhelm him. No. Relax. He had already spent too much time vainly hoping to see them come back, to embrace them once again. Until he had finally managed to come to terms with it. No. Enough of wallowing in painful hope.

But he couldn't help seeing them again in his mind's eye, just as he had known them. And his throat tightened. And tears welled in his eyes. But without rolling down his cheeks.

He hadn't been especially shocked. Why? Perhaps because other emotions had been more powerful? He hadn't had time to be shocked. That message accused him, or so it seemed. And everything else was shoved into the background.

Or perhaps this revelation, more than being a shock, had enabled him, more or less consciously, to explain a number of things to himself. At least some of the pieces of the puzzle had fallen into place. And some questions had automatically been answered. Even though, as is always the case, every new answer led to new questions.

Now he knew what his parents did, always so far from home. And how important, and secret, their scientific research was. But how, and why, were they involved with the Thaymites?

It was equally clear to him—though some points were still obscure—that his part in this whole thing was strictly dependent on his parents' past. And that the fact that he had been chosen for this mission was anything but an accident.

The professor and the counselor finally returned.

Helias regarded them triumphantly.

"When do we leave?" he said.

A tear of joy and a smile illuminated Kathia's face.

For the first time, the counselor's gaunt face was also wearing a smile.

"Good. I see that we've probably all reached the same conclusion. Excuse us for the somewhat brutal methods. But the only way we knew of to get you to see intuitively what we felt we had deduced was to trigger a strong emotion in you. It was the only way, and we needed your confirmation."

"Let me try to understand this better. Three years ago, when you found that message, you immediately started tailing me?"

"Yes, we started watching you then. First for a short period, just to make sure you hadn't done it, or at least not yet. Then, more or less sporadically, and with different people, taking turns, to make sure nothing happened to you, that you weren't approached by the wrong people or, above all, that you didn't disappear too. And finally Kathia took over. And the rest, more or less, you know."

"Let me ask a couple of personal questions…."

"Go ahead...."

"What became of my parents? And Kathia?"

"About your parents, unfortunately, we haven't been able to find out anything else. But the search continues, even though hopes are fading."

"And Kathia?"

"You can see her whenever you like." said the professor, beaming. "In fact, counselor, didn't we have something the two of us had to get done, back in my office? A question of five or ten minutes, say? Afterwards they could join us there, couldn't they?"

The two men made off. A sliding door, so well camouflaged that it had escaped his notice, opened at the bottom of the room. A couple of seconds, and Kathia appeared. Helias jumped up and went to her. But not before bruising his knee slightly when he banged it against the corner of the table.

"When do we leave?" repeated Helias Kadler.

"This very evening." answered the professor.

"There's one thing I don't quite get. Haven't you always told me that you had a veto against interfering with the past? So when did you change your mind?"

"In reality nothing's changed. The veto is still there." answered the counselor pensively.

"And so?"

"And so nothing. There's no way of preventing what we've already done, or rather, what we are now choosing to do. The stakes are too high, in any case; we have no other options."

"Are you thinking of asking for a special permit?"

"No. Special permits do not exist. There's nothing of the sort."

"And then?"

"And then it's easy. We'll do it secretly."

"Ah!"

Helias weighed this in his mind for a while. Then he continued.

"Why do you think my mother wrote that message?"

"The most obvious answer is: to leave a trace, a sign of what happened."

"That's it? Could she have been trying to urge us to go, as in fact is happening? Without that message we would not be preparing to leave now."

"I see." replied the professor. "It would be a concrete version of the 'Now you talk' that we know so well. To tell the truth, though, that seems to me to be overestimating your mother a bit, don't you think? Especially considering the anxiety and stress she was going through, no?"

"Hard to overestimate my mother. Particularly in stressful moments, when she was even more clear-thinking than usual."

The professor glanced slyly at Helias.

"So that could be who you got it from.... Hmm, possible. Maybe not probable, but possible. 'My son was here.' meaning, 'So have him come here.'...."

"And, 'My husband is wounded' could mean 'Come save him'."

"Yes. We already thought of that. Whatever your mother's real intentions might have been."

"Do we know anything else about what we're going there to do, or rather, what we've done?"

"Nothing more than vague assumptions. Which in any case might be best ignored. They might turn out to be misleading. We'll probably need all your capacity for intuition, and any assumptions we make now could interfere with it. A blank slate, my boy. We'll leave here well equipped. That's for certain. Nothing else is."

"So you are coming too…."

"Well, I know that I probably won't be any use. But an experience like this… why, I wouldn't miss it for anything in the world."

"Is there a specific reason for leaving this evening? And a reason why you hid things from me until the last minute?"

"I'm willing to bet you've already figured out the answer to the first question by yourself. As for the second, I'd prefer not to answer, at least for now."

It was Kathia who responded.

"Currently, we only have one route available for going back there at that time. If we were to leave now for the solar system, with back-now we'd arrive twenty-six days after our departure from Earth. But we have to get there one thousand one hundred and fifty-four days before, which corresponds to a round trip between here and Thaýma, always with back-now, obviously."

"So we leave this evening, we go to Thaýma and come back here some one thousand one hundred-odd days ago and, with the first available back-now, we end up on Earth in that period. Have you calculated everything in detail? Waits and downtime included?"

"We have to keep waits and downtime to a minimum because we can't afford to stay out too long on our own, given that we won't be able to take on supplies between one transmission and the other, since not only are we going backwards, but we're also doing it in secret. We won't go to Earth directly, as there weren't any transmissions to Alkenia around that twenty-third of May. Instead, we'll go straight to the Martian station where it all happened, and where there was a transmission to Alkenia that day, which we'll be able to use for our back-now down there. Mars, though, has no labs for time inversion, which we'll have to go through in order to interact with events and do our bit, as well as to be able to come back up here to our own time. Consequently, once we've got to the Martian station, we'll have to go to Earth, be inverted, and return to Mars."

"But aren't the stations all carefully monitored by you Thaymites? How can there be clandestine back-now trips?"

"As you know, back-now consists of swapping the traveler for part of the antimatter needed for transmission. In other words, whereas a normal transmission, without back-now, uses a mass of antimatter equal to the load to be transmitted, with back-now we sneak ourselves in as a portion of the antimatter, and so the release of antimatter measured at the station is less than expected. Actually, though, this release is seen as a less-than-expected acquisition of antimatter by the receiving station, the station where we sneak in, and as a defective release at the transmitting station, where we go out. A back-now can't be prevented, unless there's a tip-off beforehand. Normally, the missing masses are noted at the end, and that's about it."

"And so our trips should have left a trace. Have you checked?"

"For the first leg, from here to Thaýma, there's obviously not anything yet. The trip from Thaýma to Alkenia is absolutely forbidden with back-now, like that from Earth to Alkenia, for the same reasons. So it will be something worse than a clandestine back-now. We have an accomplice at the Thaýma station. His job is to release the magnetic field for a few instants half a minute before the transmission to Alkenia that we'll use today for our back-now down there. As soon as we arrive, then, we'll stick close to the magnetic ring, letting thirty seconds pass, backward in time, and we'll jump back into the ring, while an ad hoc ballast load will enable us to reach Alkenia in the past. Both stations will record a series of anomalies…."

"Can you access this information? I mean, what took place on Alkenia at that time?"

Kathia turned to the counselor.

"Have you finally been able to find out about that?"

"No. It's confidential, protected information, for obvious reasons. And I couldn't even insist without rousing suspicion. For the Alkenia-Mars leg, the situation's a bit different. The station was at the experimental stage and the transmission records would have been readily accessible, especially for Professor Nudeliev. Except that everything was seized after the business of the disappearances. In any case, the essential thing is that, from your mother's message, we know that at least the trip out will not run into problems."

"Yeah. I wonder why I hadn't thought about the trip back yet? Maybe because it shouldn't be in any way illegal, at least in terms of the time direction. Or am I wrong? Or maybe it'll be even more complicated? I don't know enough about it…."

"We have a few plans for that too, with a fair amount of flexibility. After all, we don't have any specific deadline to meet for our return. Nobody's waiting for us, in the future, not yet, nor do we know when we'll get back." said the professor.

"Or if we'll get back." added Helias.

"I'd say we'd better get ready now." said Kathia, looking at the time. "We'll be leaving Alkenia in five hours. Or rather, if everything has gone according to plan, we left exactly half an hour ago."

"Meaning?"

"Meaning that now it's almost five o'clock, at ten past seven we'll invert and leave, and at four twenty-four, half an hour ago, 'we'll take' the back-now for Thaýma."

"So we have over two and a half hours of travel to reach the station. I suppose it's not the same station where we were transmitted from Earth."

"Obviously. Earthlings must not know about the transmissions from Thaýma. They're received at the Scientific Station, staffed entirely by Thaymites. The trip takes a little over two hours, but we prefer to leave a little late, so as not to risk arriving early and missing the jolt."

"What would happen in that case?"

"Nothing good. The only labs for the opposite time inversion are on Thaýma and on Earth."

"And when is the next, or I mean the previous, chance at back-now?"

"Five days ago. But our reserve of energy would last much less."

"Ah! Sorry, but couldn't we leave a little later? Just enough to make sure we arrive in time?"

Chapter 8
The Professor, with His Most Blissful Expression

The professor, with his most blissful expression, continued to repeat "Fantastic! Fantastic!". His eyes were sparkling. He looked like a little boy about to be given the keys to his very own amusement park. If his legs had been shorter, he probably would have been swinging them back and forth beneath his seat.

"Fantastic! Fantastic!"

Helias, by contrast, didn't see anything fantastic about it. In fact, he was even a bit disappointed. They were in a small dimly lit wagon running along a rail of some kind, down an endless tunnel. Endless and almost entirely dark, with occasional patches of faint light punctuating lengthy stretches of gloom.

He didn't like the coveralls either. They made him feel clumsy. One size smaller would have been better, he thought. Kathia's, on the other hand, were maybe a bit too tight. Drew a little too much attention to her figure, if you asked him. Both down below and up top. Particularly up top.

And then there was that other guy, whatever his name was, who looked like he'd just stepped out of a fashion magazine or something. As if he'd been born with those coveralls on. Like a second skin, grown along with him. They should have given him a larger size. And one size smaller for himself. Maybe he'd had it made to measure, what with the job he did. And Kathia was looking at him. And why was Kathia looking at him now?

Kathia started giggling, she'd followed his whole train of thought. She turned toward Helias, goggle-eyed with mock admiration as she slowly ogled him up and down. Then a long, low whistle of appreciation. They both burst out laughing, while the others turned to look.

The wagon started to slow and the tunnel began to brighten.

"Did you pass it on to Mattheus?" asked Kathia.

"Yes. Together with the key to my room." thought Helias.

Now they had come to a standstill and the doors opened. They all already knew what the professor, who had been waiting in front of the door for some time, would have said as soon as he set foot on the ground.

M. Villata, *The Dark Arrow of Time*, Science and Fiction,
https://doi.org/10.1007/978-3-319-67486-5_8

They were in an enormous cave, lit nearly as bright as day. Part of it seemed to have been dug into the mountain, probably starting from a natural cave, while part was built outwards, ending with two gigantic doors spaced around forty meters apart.

"We're on the south side of the mountain. Those two apertures open onto a narrow gorge that makes them invisible from outside. And then everything is also very well camouflaged." explained the professor.

A small shuttle was parked a bit farther down, by the nearer door. Much of the cave was occupied by a sort of semicircular track protected by a transparent tunnel.

A few people, apparently staffmembers, were in some kind of control room beyond the semicircle. Others were coming and going, more or less busily.

"Fantastic!" murmured the professor, again.

Now the bustle had quieted. Only two people, probably carrying out the final checks on and around the shuttle, were still moving. Then they too left. It seemed that everything was ready.

The pilot—that fashion victim, whatever his name was—was talking on his cell, probably with the control room. Breaking off the conversation, he motioned to the others to follow him, and walked toward the shuttle.

It was a six-seater, and Helias understood why there would be six of them in the departing group. In addition to himself, Kathia, the professor and the pilot, there were two 'chaperons', probably a security detail of sorts. One was tall and swarthy, with a heavy jaw and a scar down his chin, ill-concealed by thick stubble. The other was thin, dark blond—though his hair seemed inclined to beat a retreat—and a bit older. Neither said a word, limiting themselves to minimal movements: latching their seat belts; putting on their helmets.

The pilot started the engines and the shuttle lifted off the ground. The countdown started on a display. 148, 147, 146.

"They've already told you what will happen shortly, I imagine." remarked the professor, leaning towards Helias's ear from his position behind him.

"More or less." answered Helias, twisting around in his seat. As he did so, he caught Kathia's eye, as she sat alongside him in the row to the left, less than a meter away. Apparently his face betrayed a certain unease, since Kathia was watching him steadily and smiling, as if to calm him.

"Are you afraid, young man?"

"Shouldn't I be?"

"Not at all! Everything will be just fine, you'll see."

124, 123, 122.

Meanwhile the other door, at their left, had slowly opened and a shuttle similar to theirs appeared.

"Things'll be fine. Guaranteed." repeated the professor as he watched the other shuttle taxi in.

"Only remember, my boy, that you mustn't let anything frighten you. Whatever happens. Or you think is happening."

"What do you mean?"

"You'll see in a little while. But why do you ask? Didn't they explain it to you?"

"Explain what?"

"There's not enough time now. We're almost ready to leave."

99, 98, 97.

The overhead lights and those in the shuttle had dimmed, and now only the inside of the transparent tunnel was illuminated as they entered.

"Magnetic fields activated." said the pilot.

"Did they tell you about the two membranes?" continued the professor.

"Yes, something was mentioned."

"Now they're being charged. They are barely a couple of millimeters apart. But they have to be crossed at very high speed. The first subtracts the first half of your energy, and in those two millimeters time no longer passes for your body. The second membrane takes the other half from you, and your time will be inverted. As you are going out of the second membrane, you belong to the other arrow of time. That's why you have to cross them quickly. It can't be much fun to prolong a situation where your heart, or a lung, or even your brain, is in slices, with each going its own way, can it? Fantastic! Don't you think?"

The professor was clearly speaking more for himself than for Helias. Like a little boy gazing at his ice cream in happy anticipation before starting on it in earnest.

"So you can expect a violent acceleration slightly before crossing the membranes. Maybe six or seven seconds before. But remember, no panic."

"Have you done this before?"

"No."

"But I have." interjected Kathia with her most reassuring look.

"Helmets."

18, 17, 16.

"Pressurization."

They were now nearing the end of their quarter-circle and, around the curve, could glimpse the translucent membranes.

As he looked in that direction, however, Helias realized that another shuttle, in the other half of the transparent tunnel, was moving toward the membranes. And toward them. It must be the one he had seen taxiing a little while before. But he had stopped watching it to listen to the professor's explanations.

"What do they think they're doing?" he couldn't help but mutter inside his helmet.

10, 9, 8.

"Acceleration." said the pilot.

"But what are they doing? Are they accelerating too? They'll collide with us!" shouted Helias.

"No panic." repeated the professor.

"It's just us, going away." explained Kathia in reassuring tones, as she grasped and squeezed his hand.

Helias had no time to say or do anything else, apart from instinctively bracing his feet. Beyond the translucent membrane he could see the other shuttle hurtling toward them. But there was no crash. As the two shuttles reached the membranes simultaneously, they disappeared layer after layer, as if they were cancelling each other out. The 'impact' lasted the barest instant. And then Helias found himself on the other side of the membranes, together with the others, a few moments before the impact. Still with the membranes ahead of him. But this time they were moving away, backwards. As if they had bounced back, rather than passing to the other side.

The control room had seen the two shuttles collide and annihilate each other. Leaving no trace. Except for the enormous amount of energy absorbed by the membranes and now stored elsewhere. The shuttle and its passengers no longer existed. But they belonged to the past and were going backwards in time, exactly where they had been seen to pass, in the other part of the semicircle.

"Inversion." the pilot had said. And the cave had spun over, floor down. And the membranes were now moving away behind their back.

Helias was in an adrenaline rush, and realized that he was still fiercely clutching Kathia's hand in his.

"Here we are. Fantastic!" exclaimed the professor.

"And there we are, slightly before starting to accelerate. Except that now we're moving away from the membranes. Like us, in fact. Here and there simultaneously. But 'we' are a little older. And so we remember being there.... Fantastic!"

"But why did we come out backwards? We went in face forward and came out back to front. Why?"

"Because a time inversion also entails a space inversion, otherwise the transformation would be improper. The Lorentz total inversion, remember? And why were you frightened when you saw the other craft? Don't you remember the Feynman diagrams? And yet I explained them to you, as I recall...." he said as he took off his helmet.

"Eh, it's one thing having them explained to you or even thinking about them. Living inside them is another thing altogether. And then, as you will remember, I also fell asleep that time, when you were explaining.... Professor, why are you taking off your helmet?"

"Because what I didn't remember is that you have to travel on an empty stomach. Or maybe I didn't think it mattered. Where is the bag?"

The big doors began to open. Helias turned to the porthole to look at the other shuttle. It had just left, or in other words was coming out of the tunnel as he watched, while the outside lights and those inside it brightened. For a fleeting instant, he was able to make out the faces of the people on board. And he couldn't restrain himself from putting his thumbs in his ears—or where his ears would have been if they weren't covered by the helmet—and fluttering his fingers at the professor, who was looking towards him from the other shuttle. Satisfied, he looked ahead, down the narrow gorge at a sliver of starry sky, repeating to himself, "Fantastic! Fantastic!".

The professor had just come back from the lavatory, looking distinctly unsettled.

They had come out of the gorge and were flying over the castle and lake.

It was suppertime, and there were only a few scattered people going about their business around the castle. Walking, obviously, backwards. While the waves, yellow and orange ripples reflecting the light from the castle, lapped outward, away from the shore.

"Everything okay, professor?"

"Yes. Much better. Thanks."

"How is it that we're flying over the castle? Are we determined to attract attention to ourselves?"

The professor didn't seem to be quite back to normal yet. It was Kathia who answered.

"They can't see us. The magnetic field that surrounds and isolates us can be configured to block our advanced radiation, which is what would make us visible in the other time direction. Likewise, we're transparent to the retarded radiation of the matter around us, and so we're perfectly invisible to the other arrow of time."

Helias seemed doubtful. The professor chipped in.

"As I've already told you, we all can intercept, and thus see, only the radiation coming from our own past. And so we now see ordinary matter thanks to its advanced radiation, to which, however, that matter is transparent, as it comes from its future. In other words, we can see 'all' the advanced radiation, except for any which might have been absorbed by 'antimatter'. For me, in fact, you and the entire craft are not transparent. Nevertheless, we should be able to 'see' through the walls of the building below us, for example, or the starry sky through the mountains, or even the sun through the planet, and so forth. But no, we can't, it wouldn't be practical. We'd run the risk now of crashing into that dark mountain, and we'd be perennially disoriented and confused, like in a world made of glass. The fact is, that even if advanced radiation doesn't interact with matter, crossing opaque matter changes its polarization, and this becomes a selective means of reconstructing a 'normal' image. This glass in the shuttle portholes filter the appropriate polarizations so that we don't perceive light as total chaos. Would you like a demonstration? Pilot, excuse me, could you please go back down towards the building again and dial back the filter effect?"

The shuttle reversed its course and approached the castle, hovering in midair. Gradually the outer walls and windows became transparent, as did the objects and people beyond the walls, and the floors and secondary walls. Like stained glass, its colors layered atop those behind, dimmer and dimmer as the distance and the intervening mass increased. Or like patchy reflections in a shop window, where the darker spaces still afford glimpses of the interior. From their vantage point, they could see people climbing the stairs backwards, probably on their way to the dining hall. A shower was running, the spray rising and climbing back into the nozzles, a girl coming out of the stall with her hair completely dry. And Kathia's interposed face, winking at him.

Just as gradually, the walls became opaque once more, and the shuttle resumed its trip.

"How is it that when we were in the cave, before being inverted, we could see the shuttle—ourselves, that is—going into the past?"

"Because we hadn't blocked out our advanced radiation yet. Otherwise we wouldn't have seen ourselves 'arriving' and 'coming in', and so we wouldn't have been able to 'open' and 'close' the doors, or vice versa, if you prefer."

"So we saw the shuttle's advanced radiation. But shouldn't we also have seen the retarded radiation of the matter behind it? Like a sort of ghosting?"

"It depends on the relative intensity of the light in the background. In our case, the craft was well lit, the background less so, particularly when the cave lights were dimmed and the craft was illuminated by the light from the tunnel. You could see it before, though. If you had been paying attention, you would have seen the shaded areas look practically transparent, as if the craft were a bright reflection in a window."

"Shouldn't we have seen the interior of the shuttle too? In other words, its advanced radiation that wasn't intercepted by the fuselage?"

"No. For two reasons. The interior lights were very dim, so we would have seen hardly anything anyway. In addition, we were already in the craft, with the filter windows."

Helias felt that he needed to rest and relax a little. He laid his head on the back of his seat and, eyes half closed, felt again for Kathia's hand, squeezing it gently.

He could see the subdued lights in the cabin through his eyelashes. He turned his head a little to look outside. They were gaining altitude over a broad valley, ready to crest the mountains at its edge, their snowy peaks glazed with starlight. Light from Nasymil, rising on the horizon.

The professor seemed to be his old self again. He was talking in his helmet, without knowing whether anyone was listening or not. As if he were giving a speech to an inattentive audience. Talking about the light from the stars and the remote galaxies. About how the farther away they were, the more you could see into their distant future. The opposite of what happens for the normal astronomer, who observes their distant past. And he hoped that there would soon be detectors made of antimatter that would be able to provide details of the universe's future which, together with our knowledge of its past, would make it possible to answer many of the open questions of cosmology and astrophysics. Indeed, such discoveries could not fail to lead to a sweeping revolution in these fields of science, the likes of which had perhaps never been seen before.

The professor continued with increasingly incomprehensible examples, something about the evolution of active galactic nuclei, distinguishing between—or equating, it really wasn't at all clear—radio sources, Seyfert galaxies, quasars, blazars, and something to do with the constellation Lacerta. But Helias had stopped listening. He was attracted by that glow on the horizon, now reflected in the waters of a river that was flowing back to the spring. Part of the river branched off toward the left, and further on he could see the waters merging, collecting together in a pool as if to gather the energy needed for the breathtaking leap back to the top of the cataract.

Soon the sun rose. Just as it was setting for the 'others'. But Helias knew that 'his' sun, the one he saw, was a little older.

Older, but growing younger.

They were climbing. Faster and faster.

Helias was extremely uncomfortable. Not because of the acceleration, he didn't even notice that. It was seeing the planet fall away beneath them, so quickly. It was that overwhelming nostalgia for good, solid ground under your feet. It was no longer knowing 'up' from 'down', and your mind can't keep track. He had never done anything like this before, never had what liftoff really means brought so forcefully home to him, with the planet plummeting away. And a mixture of intellectual euphoria and raw, atavic terror coursed through him. He was also very, very dizzy. Except for the transmission from the Earth to Alkenia—but he had slept through that—his only other experience of the kind had been, years ago, his departure for Mars. Then, though, there was no view of the liftoff from the spaceship.

They arrived at the Scientific Station well ahead of time, all a little dazed after the trip, some perhaps drowsy. The pilot urged the passengers to rest, and the seats tilted backwards. The professor asked to be woken a bit before the transmission. Kathia and Helias did the same. The countdown on the display showed slightly less than an hour.

Only four minutes to go. The engines started again and the shuttle proceeded toward the fluorescent loop. The Thaymite cargo ship, just out, moved backwards toward the loop. They had to enter together, less than a fraction of a second apart, at different angles and converging in the loop a few meters away.

The pilot ordered silence, though none of them dared speak anyway. A device was determining the necessary triangulations, which appeared on a screen. When the shuttle was within a few centimeters of the exact relative position, the pilot activated the automatic stage: from then on, the shuttle would maintain the triangulation however the cargo ship varied its speed. Helias found himself thinking that all this had already happened, in the other time direction: so why all this precision and pinpoint timing? But the answer was obvious: so that everything would happen as it had already happened.

A few seconds. A few meters. Again, the feeling was that they were going to crash against the 'other', just as they had felt before, in the tunnel with the membranes. But the magnetic field had opened, unleashing a violent rain of plasma.

Then nothing. As if nothing had happened, the cargo ship was now moving away, backwards, on their right. As if they had passed straight through each other, unwaveringly, each maintaining its own energy and its own route.

But it wasn't the same station anymore. And that green, white and blue planet down there was not Alkenia, but Thaýma.

Not even the time to admire the scenery. A new countdown had begun. And was already about to end. A rapid reversal of route. An acceleration and another plunge. Without even time to see what the counterpart 'coming' from Alkenia might be.

And there was Alkenia again, three years earlier. But no counterpart in sight. Helias asked the professor to explain.

"I don't know, my boy. Perhaps, simply, we can't see whatever it is because we're now in Alkenia's cone of shadow and consequently it's not illuminated by the sun."

"But there's the light from Nasymil. Look at the station, you can see it quite well."

"The station is big. Our object could be small and far away by now, or it could simply be black and opaque. Or against the light from Nasymil."

"But we didn't see it before, either, coming in."

"Well, everything happened so quickly that I might not have seen it even if it was an enormous, coruscating ham sandwich. Maybe we were looking in the wrong place. Or maybe it was below us. Did anybody see anything?"

A chorus of 'nos' echoed around him.

"Pilot?"

"My job was to follow the countdown and bring the ship into the center of the loop. Those were my only orders. Perhaps, as you say, it's just that the object was black. Besides, it's reasonable to think that it was invisible to everyone…."

"Wait a minute!" burst out Helias. "Whether it was black or not, who was it who prepared this counterpart, or 'ballast', as you called it? What an idiot I was not to think of it before! It can't have been prepared over there, on Thaýma, and then sent back in time, like us. It must have departed from here and appeared over there later. Not the other way round. But who can have done it? Who, here, could have known, three years before?"

A short hum of murmured comments. Then silence. As if they were waiting for an answer. The professor, craftily, smiled.

Helias smiled too. "Us! We're here, three years before. But what can we do? How…? Professor?"

The smile still on his lips, the professor whispered "We'll see…. We have plenty of time. Now we have other things to do. We'll see…."

They flew over Alkenia. And had something to eat. There were still hours to go before the rendezvous. The pilot recommended a little dose of sleep spray and a nice nap.

Then a station again. Triangulations. And another collision that didn't happen with a little spaceship, this time coming from Mars.

And there was Mars, reddish. And the station. More or less as he remembered it. Binoculars. A little walk around, a tour of inspection so that Helias could recognize the places he had seen with his parents years before. They had a map, which they had studied before the last 'jolt'. Better, though, to have somebody who already knew his way around the station, much better. So a walk around would help refresh his memory.

Their 'counterpart' was docking now, or in other words was casting off from the Martian station. Approach. Zoom in with the binoculars. Someone moves quickly, backwards, along the short narrow passage that leads to the docked ship. Others follow. Two people, maybe more. It is hard to count them and see who they are through the small portholes along the passage. Nobody now. Further on, though, after that curve, there's a large glassed-in area where they might appear in a moment. Maximum zoom. There they are! Too fast to make them out clearly. One, two, three. Then a fourth. Who stops for a second. Turns around, long hair bouncing. Then disappears, like the others. And the light goes out.

They were all struck speechless. Especially Kathia. Whom everyone had recognized.

A light a little farther on. A big window giving a view of an apartment. A handful of people enter backwards, maybe the same ones as before. They are far from the window, and can be seen only from the chest downwards. A woman comes in, back first, but from another door. She turns and embraces a man, who then leans toward the floor, where something that appears to be a body stretched full length can be made out. As can Kathia's blonde head, next to someone else. A man comes and goes from the door where the woman 'entered'. Then Kathia, too, bends over the body on the floor. She puts something in a backpack, rises and, backwards, goes out the door, preceded by two people. A man and a woman remain next to the inert body. Then they rise too and go through the same door, leaving the room empty.

This time it was Helias who was dumbstruck. That woman who had now disappeared beyond the door, and who the man—who he!—had embraced, was his mother. No sign of his father. But it was likely that the half-concealed body sprawled on the floor was his.

Then the lights went back on in the corridor, and once again there seemed to be movement in the narrow passage leading to the ship, followed by the appearance—showcased, almost, in the large glass pane—of a portly gentleman who was trying not over-gracefully to hasten down the corridor, and whom they all thought they recognized.

This was when the professor intervened. The professor on the shuttle.

"I suggest we stop this. Stop staying here and spying on ourselves, I mean. Aside from the fact that watching these people go back and forth backwards is giving me a headache, I'd say it's psychically healthier to stop viewing this. Otherwise, what kind of mental confusion is going to overcome us when we find ourselves there, going through these gestures again? And I hardly think it's a good thing, in general, to be spectators of our own selves."

"I agree." said Kathia. Helias nodded too.

In reality, Helias had been about to ask if it was possible to turn down the filter effect so that they could try to see what was going on in the adjacent rooms, on the other side of the door they had all disappeared through. But he quickly backtracked upon hearing the professor's words, and decided to say nothing.

First to speak was 'Scarface' (so Helias had nicknamed him, privately, since he didn't remember what he was called; the other heavy was 'Skullet'). It was perhaps the first time he had heard his voice.

"I agree too. Particularly because, seeing as we're already there, that means we'd better get a move on, wouldn't you say?"

He couldn't wait to have a chance to kick a little ass, evidently.

The professor looked him up and down, condescendingly. As if he was seeing him for the first time; as if to say, 'And you would have been better off keeping your mouth shut, wouldn't you say?'. Then he gave a sigh and, eyeing him steadily from under his brows, answered slowly and deliberately, almost spelling out the words.

"First: if we're already there, what need is there for us to get a move on? What's the hurry? Second: the more we hurry, the less time we'll have, since now our time is running backwards, remember? Instead, if we take our time now, before our inversion, I mean, we'll go further back in the past, and so then we'll have more time to return here now. Don't you think? Do you understand, now?"

Sullenly, Scarface clammed back up.

And so they went towards Earth, not hurrying.

The time inversion station was in orbit around the Moon, camouflaged as an ordinary space lab.

They were almost at the limit of their range. By now, they had to invert, so they could take on supplies.

Kathia explained to Helias that the opposite time inversion was a more delicate matter. The lab consisted of matter going in the time direction opposite to theirs, unlike in the previous inversion. When they had inverted on Alkenia, there hadn't been any problems in communicating: both they and the lab came from the same past, and so the inversion process could be triggered because they were there, in the same past as the inversion. Now it was different. They were in the lab's future, in the inversion process's future. And so they couldn't announce, in the normal way, that they were present and ready to be inverted. The station knew nothing, couldn't know anything, and would know only when it was all over. Suddenly, out of the void, the two components would have appeared, both visible and existing only in the inversion's future, whence no signal could have been received.

Unless there were an antimatter device that, belonging to the same arrow of time, would have no trouble detecting their presence in its past. So that the inversion process could be started.

At a certain point the pilot said, "We're ready, do you want to proceed?".

The professor gestured toward Kathia, saying, "Please, you do it, your pronunciation is better than mine.".

Kathia took off her helmet and approached the cockpit. She picked up a sort of microphone and looked for the switch, asking the pilot for confirmation. Then, in a language Helias didn't recognize, she said a few words.

At the end of the phrase, a two hundred and forty second countdown began. Kathia handed back the microphone and returned to her seat.

"Helmets and seat belts." said the pilot.

"What language was that? What did you say?" asked Helias.

"It's Ancient Greek, like the word 'Thaýma', remember? I said '῞Ηκομεν γάρ ἐκ τοῦ πλανήτου Θαύμα.', which means 'We come from Thaýma', more or less. It's our password. To trigger the inversion process."

They approached the space lab, toward the big hatch on the right.

A ship appeared from behind the station, on their left, going rigorously backward.

"There we are." said the professor. "We're going to take on supplies, on the other side of the station."

Slowly, the hatches began to open. And the two ships moved in, all according to plan. As before, a semicircular track, in a broad transparent tube. Beyond the big glass panes at the end, a few people, surprised by the arrival, waved their arms in greeting. Helias, euphoric, waved as well, now toward the windows, now toward the other shuttle, where another Helias waved back. And another professor, one who hadn't had his lunch yet, made faces at him. Both Kathias smiled. Scarface and Skullet, not to be outdone, sketched a handwave of sorts.

And then the acceleration. Ka-Boom. And the roles had inverted.

Inversion. Still a few waves, here and there. And then away, for supplies.

Chapter 9
As They Came Closer to Mars

As they came closer to Mars, Helias got tenser and tenser. And more withdrawn. One of the great events in his life was about to happen. And, like Mars, it was coming ever closer. He had stopped talking. Kathia watched him, gently stroking his arm. As if to keep his courage up, but without intruding, without distracting him from his thoughts, his musings.

Kathia followed his meandering thoughts. Painful, crowded thoughts. But hopeful ones too. Full of hope that, bit by bit, bloomed into optimism, almost euphoric. Kathia slid her hand toward Helias's, ending palm to palm.

"Professor. We're about there." she whispered from inside her helmet.

The professor pulled out his pocket computer and handed Helias a small oval object.

"Keep it in the palm of your hand, please."

Helias turned questioningly to Kathia.

"What is it?"

"Don't worry. I'll explain later." interjected the professor.

Then, looking at the computer, he added "Intense. Between eight and nine. With regular interference beats.".

Helias let it rest there, and went back to his thoughts.

They had almost arrived. Mars was looming over them on the left, and he thought he could already make out the station straight ahead. The shuttle slowed, and the pilot announced that they would be docking in a few minutes.

With practiced hands, Scarface and Skullet loaded and inspected their weapons, a bit bigger but probably the same type as the ones Helias and the others had. Maybe they emitted more powerful and more accurate paralyzing beams.

A flyover with the lights off, binoculars trained on the station. To see what, if anything, was going on.

The station seemed strangely deserted. Three ships of varying sizes were docked at the boarding bridges. Two of the bridges gave onto the main corridor. The third ship was the one they'd already seen, and that they would be leaving on later.

M. Villata, *The Dark Arrow of Time*, Science and Fiction,
https://doi.org/10.1007/978-3-319-67486-5_9

They already knew where they had to go. The problem was getting there. The lights were on in the room in question, but the windows were opaque.

They approached the bridges on the main corridor. Through the windows they could see several people, stationed as if to control the entrances. They were armed.

Helias remembered a secondary entrance. A little boarding bridge that led directly to a lab, and used for unloading materials for it.

They docked and went into the lab. Deserted. Maybe all the workers at the station had been rounded up and locked away somewhere, prisoners.

In silence, they threaded their way down the narrow passage that led to the main corridor.

They kept careful watch on the guards, moving about a few dozen meters farther down. At a certain point only one man remained in view, his back to them. They took advantage of it to sneak across the main corridor and into the smaller hallway that went to Helias's parents' lab and living quarters. Skullet led the way and Scarface brought up the rear of their little line. Once in front of the lab, they heard voices from beyond the door, which was ajar. There were two men. One was busy at the main computer console, the other was talking on his cell, probably with the guards.

"No. He wouldn't cooperate. Not at all. We had to neutralize him, in fact. Yes.... No.... We're not getting anywhere yet.... He's trying.... No, stay where you are. And double the guard on the other prisoners, there might be other hotheads. No.... Here everything's under control, but we're not getting anywhere. We need reinforcements, specialists. Experts in this kind of operation. It's all more complicated than expected.... Yes, do that. Make contact from the ship.... Yes. Over and out."

They had listened to everything with attention. Skullet looked at the pilot and jerked his head toward the door. The pilot nodded. Skullet and Scarface stood on each side of the door, locking eyes. Without warning, they burst through the door and, one on his feet and the other on the ground, opened fire. The two men, hit in the arms and legs, crumpled and collapsed. They were dragged into a corner and bound and gagged with tape. It was all over in a question of seconds.

Skullet picked up the cell that had started to ring and clamped it to his ear. He answered, trying to imitate its owner's voice.

"Yeah.... Okay.... We're trying another approach. We need to concentrate. Call only if necessary. Yes.... Fine.... Over and out."

Kathia, who had been the first to enter the lab, was at her most serious, furrowing her brows.

The professor had gone to the console and was looking at the screen.

The pilot, turning to their two defenders, had said "You two keep an eye on the corridor.".

Helias had headed immediately to the door at the far end, the door to the living quarters. Kathia had followed him after a few moments.

Helias's mother was in tears, kneeling next to her husband's prostrate body. She was trying to stanch his wounds with anything, everything, that came to hand.

"Mama!"

"Helias!?"

"Yes, it's me, but there's no time to explain." he said, holding her hand in his and looking her in the eye.

"But it's not really you." she said, stroking his hair. "You've come from the future? How?"

"Yes, mama. In a sense you called us yourself."

She didn't understand. In the meantime Kathia and Helias had bent over the supine body.

"Damn them! They used invasive weapons. He's losing blood. From the leg and head." said Helias as he took his father's hand, which remained limp.

Kathia was feeling his neck for a pulse and raising his eyelids.

"Bradycardia. Fixed dilated pupil. Homolateral mydriasis. We have to stop the hemorrhage in the leg."

She had taken off the backpack she had brought from the shuttle. She took a flask, a scalpel, a bandage and a tourniquet from it.

"Ma'am, while we're dealing with the rest, free your husband's leg and tie this above the wound. Then I'll have a look."

The woman did as she was told, while Kathia drew a strange sheet from its case and, at the same time, inspected the injury to the head, above the ear.

"Done."

Kathia put on special goggles and leaned over the leg.

"Good. The bone is intact. Take this bottle and spray the area abundantly and wrap it with this, not too tight. Then remove the tourniquet."

Still wearing the goggles, she turned toward the wound on the head. With a few rapid movements she shaved the hair all around and cleaned the wound. Then she inspected it carefully.

"Comminuted skull fracture. Probable meningeal and cortical injuries. When did it happen?"

"I don't know. I think I've lost all sense of time. I'd say twenty minutes ago, half an hour at most. Is it serious?"

"Yes, but we can save him, if we hurry. The worst symptom is the fact that the midbrain is involved, probably compressed by a cerebral herniation. Now we'll have a better look. And while I'm working, tell me how it happened."

Kathia sprayed the wound, unfolded the sheet—a thin membrane—and sprayed a part of it too. Then she placed it on the man's head and cut a hole for the nose and mouth. With gentle pats, Kathia made the membrane adhere to the head and face, almost like a second skin. Kathia handed a pair of goggles to Helias, who put them on and began to see through the membrane and the skin. Kathia told him to move his eyebrows down and up, to penetrate the different layers of tissue.

Meanwhile the woman related how those men had invaded and occupied the lab, trying to convince them, and then force them, to hand over the compiler. They had refused, obviously, and the men escalated their threats. When they had pointed a weapon at the woman, the man went ballistic, throwing himself against the aggressors. First they shot him in the leg. But that wasn't enough to stop him, and so they toppled him with a crushing blow to the skull. When they threatened to finish her husband off if she didn't talk, the woman, as her only way out, pretended to faint. Then they had dragged them both into the living quarters and locked them in.

"Animals!" Helias had burst out.

His mother was looking a little less distraught now. She had someone to share responsibility with. She had her son. Miraculously come to their aid. And there was an expert physician with her husband, and maybe she would be able to save him.

Helias was practically shell-shocked. By the fact that his mother was there, looking at him now with gratitude. His father was there, unconscious, in a congealing pool of blood. But alive! And so was his mother! Now he had to do everything he could so that they would keep on living. And then there was the amazement for this new version of Kathia. Kathia the skilled physician, whose precise gestures and professional concentration he was admiring. And who was saving his father's life. He finally managed to stammer, "You, you're a doctor?".

Kathia, who had followed some of his thoughts, smiled.

"You, what would you say?"

She had made two small holes in the membrane, inserting two rods, one with a sharp blade, like a tiny scalpel, that she was handling with a practiced dexterity.

"I've removed some bone fragments. Now I have to drain and evacuate the hematoma, and proceed with hemostasis. Then I'll remove the parts of the dural and cortical tissue that are damaged. After that I'll need your help, Helias. Get ready."

"To do what?"

"Nothing special. I'll tell you in a minute. But now go wash your hands."

Kathia made a third hole, threading in a thin catheter. The blood began to spurt and flow into the reservoir at the other end. Then she started working with the rods again, occasionally replacing them with others and interchanging them. Finally, she drew a sigh of relief and scrutinized what she had done closely. She drew out a new pair of rods, inserted them and worked a little more. Then she looked at Helias.

"Your turn now."

"Tell me what to do."

"Spray your hands and hold them around the wound, a few centimeters from the membrane."

Helias, though far from convinced, obeyed.

"You don't mean to tell me that…?"

"That what? That you're about to heal your father? Why not? Isn't that what you want?"

"Sure, but…."

"But what? You don't know how to do it? Really? And yet your hands release more energy than you could possibly imagine. And not only the right frequencies. But above all the right interference beats. That's what counts. Go on. I'm waiting. We have to reconstruct the damaged parts."

"I…. I… d-don't know…. What is it I have to do, exactly?"

"You're asking me? You're the biologist. I'm only a physician."

"What? What do you mean? I'm supposed to concentrate on the right cytology? Do a pertinent histological review? Imagine cells, but the right ones, that divide and proliferate? And where do I put the blood vessels? And who remembers what dural tissue is like?"

"No, no. I was joking. But I see that I've put you in the right mood. Look."

A pale blue light seemed to pervade the open wound beneath the membrane. Then it changed color, pulsating.

"All you've got to do is save your father. That's all. You only have to let yourself do it. Isn't that what you want? He gave you half of your life. Maybe it's time to return the favor, don't you think? Who knows his cells and their content better than you, even though you don't know that you do? Good. I see everything's clear now. Look."

While speaking, Kathia had been moving her rods, adroitly as usual.

In the meantime, the professor had entered the room too. He had adjusted the transparency of the outer windows, and now he was standing off by himself, keeping well out of the way. He shied away from looking at the scene, probably not all that thrilled to see blood and an open cranium.

The pilot appeared at the door, but went away immediately when he saw he wasn't needed.

The 'operation' lasted several minutes, during which Helias felt intense heat in his palms. It wasn't like feeling heat from some outside source, no, more internal, his hands themselves unleashing incandescent energy. And there was a tingling, too, varying in intensity like little electric shocks.

The work was almost done by now. Underneath the blue haze you could just make out the reconstructed cerebral cortex. Next the dura madre was sutured and the bone fracture was repaired, using the fragments that had been removed earlier.

Once the flaps of skin had also been joined back together, Kathia removed the membrane and covered the area with a broad bandage. Then she moved off to the side and began to put away her equipment.

Helias was exhausted, but happy. And filled with a deep peace.

His mother, who had watched the whole operation in wonderstruck silence, had now taken Kathia's place alongside her husband and, with wordless emotion, stroked his face delicately with the backs of her fingers, her other hand gripping Helias's.

Coming to, Helias's father opened his eyes. Helias smiled and a tear wet his mother's cheek. Kathia smiled too, and then rose and walked toward the professor to consult with him.

"Where am I? What happened?" asked the scientist, looking at his wife. Then, following her glance, he turned his eyes.

"Helias!?"

"Yes. He came to save us. We were attacked, remember?"

"Hi dad."

"Hi Helias. Yes, now I remember. But what are you doing here? You seem different...."

The voice quavered, and he spoke with difficulty.

"You have to rest now, stay calm. You have to recover from a nasty injury. Mama will tell you all about it. You just have to tell me where the compiler is.

We'll probably have to get out of here any minute, as soon as the guards begin to suspect something. We can't risk a face-off with them."

"Who are you with? How many of you are there?"

"There are six of us."

"Where are the others?"

"Two, or maybe three, are guarding the corridor. Then there's the professor and Kathia. Unfortunately we don't have time for introductions."

"Yes. Get out of here right away. I think there are a lot of them, and they're well-armed, as well as trained and absolutely ruthless."

He tried to move an arm, but had to give up.

"Put your hand in the side pocket of my trousers. There's a small adhesive pocket inside, hidden by the seam."

Helias did. And found what he was looking for.

"I'll be right back. I'm going to consult the others."

Kathia and the professor had joined the pilot in the lab. He was talking to the two prisoners.

"One of you two will come with us. As a prisoner and hostage. We obviously recommend that you do no harm to the two scientists, whom we can't take with us. You have nothing to gain from it. And if you do, your man will suffer the consequences. Your mission has failed. I repeat, harming them will do you no good. And there's no doubt that whoever sent you here would punish you for it. They're much more use to him safe and sound...."

Helias, meanwhile, had shown the compiler to Kathia, who took it and approached the two men, showing them the diskette.

"As you can see, there's nothing else for you to ask or look for. Anything to say?"

One of them nodded his head. The pilot removed his gag.

"We're not criminals, even if we're mercenaries and we don't think much about whether what we're ordered to do is a dirty job or not. You can trust us, we won't hurt them. I command these people. If you want a prisoner, take me."

Kathia nodded. "He's sincere." she said. "Free his legs, so he can walk."

The pilot leaned over, taking out his knife to cut the tape off the man's legs. There was a noise in the corridor and Skullet appeared.

"They called me again, alarmed at not hearing from these guys. I told them to wait, but I'm afraid they're getting suspicious."

Kathia, frowning again.

Another noise. And Scarface came in.

"They're coming. I locked the two doors between us, but it won't take them long to break through."

"Quick!" cried Kathia to the pilot.

Then she spun around and struck Skullet's hand, hard, and the weapon he had been aiming at the prisoners clattered to the floor. A twirl, and a kick, and Kathia sent Skullet sprawling.

"What are you doing? I just wanted to put them out of action. They're animals, don't you see what they did to the lab director? They don't deserve any better."

"From now on you'll follow orders and do nothing on your own initiative." answered Kathia.

The man rose, cursing, and collected his weapon.

Kathia turned to the pilot.

"You, the professor and the prisoner will precede us to the ship. Have the woman in there tell you how to get to it."

Helias, thrown completely off balance now by everything that had happened, but surprised and amazed too, managed only to blurt out, "Is there anything else you can do that I don't know about?".

Kathia looked at him and winked.

"Many things…, my boy."

A few seconds later, Helias's mother came into the lab and went up to the mainframe screen. She touched a few points with her fingers and lists appeared. She selected a file and copied it on a diskette, which she handed to Helias.

"Here's the executable for the password program. The compiler is password-protected. The password is complicated and unknown, but it's produced by this simple program after entering the correct input string, which is thus a sort of pre-password. The instructions for identifying the pre-password are on this sheet, which only Helias can easily understand. And so the compiler, to be operated, needs both this program and these instructions. I'd advise you to split responsibility for them among yourselves, to make your enemies' job harder."

Skullet approached, apparently volunteering. But Kathia, with her most serious look, took the diskette from Helias's hands and tossed it to Scarface, who was keeping an eye on the corridor from the doorway. She kept the compiler herself and told Helias "You'd better look at those instructions and memorize them. Then destroy the sheet.".

As Helias read the instructions, his mother came up to him.

Then Scarface said, "I think they've broken through the first door, we'd better get out of here right away.".

Kathia took a chair, closed the door leading to the lab and jammed the chair up against it. Turning to their two escorts, she said, "Tie the prisoner to this chair. Take the tape off his mouth. It'll gain us a few minutes, if they don't want to shoot him full of holes and roast him with their lasers.".

Helias and his mother had gone back into the living quarters and Helias was caressing his father's face.

"Bye dad. Take care of yourself."

And he winked at him, to hide a tear.

"Bye mama." he said, folding her hands in his.

Kathia came in too, followed by the other two, and she bent over the man as if to check his condition.

Helias's father stared in open-eyed amazement.

"You!? What are you doing here?"

"Dad, it's the doctor who saved your life. It's Kathia."

"Do we know each other? Perhaps you've mistaken me for someone else. I don't think I've ever met you before…."

And so saying, Kathia drew a vial from her backpack.

"Sorry, but we're in a hurry. I've got to drug you. It's much better if they find you asleep and you don't give them any reason to do you any more harm than they already have."

And without giving him time to react, she sprayed the vial in his throat.

"Kathia! What does this mean?" asked Helias, stunned.

"Nothing more than what I said." she answered, serious. "Anyway, in his condition, he needs to sleep and, above all, to avoid getting upset."

Then she rose and told Skullet, "Go to the other room and check what's going on.".

She moved to where Scarface was standing and conferred briefly with him.

In the seconds before falling asleep, the wounded man had found the strength to grip his son's arm and, fixing him in the eye, whispered "Be careful…. My son….".

Skullet had come running back.

"They're at the door."

And a voice was crying "Don't shoot! They've tied me behind the door, don't shoot….".

"Let's go!" Kathia had said.

"This way, follow the corridor." Helias's mother had indicated. "I'll open the loop for you. And I'll try to keep them from following you there…."

One last emotional embrace. One last look at the man who lay sleeping. One last look, with a lump in his throat.

Helias turned again toward his mother. He tried to say "Write….", but his voice failed him. And she, too, bid him farewell with one last look and disappeared through the doorway. But she would write anyway. He knew that already.

Now Kathia was moving down the corridor, going fast. Turning her head around, as if to check that the others were following her, as if to check positions and distances. Helias was next. The other two formed the rearguard.

After the first curve, Kathia turned again, and Helias, a few meters behind, recoiled as he caught the look on her face. Then she outstripped him, and at the next corner he found her facing him, stock-still and eyes that struck fear into his heart. Her arms were outstretched, and the pistol gripped in her meshed fingers was aimed at him.

Chapter 10
The Pilot and the Prisoner Had Reached the Shuttle

The pilot and the prisoner had reached the shuttle some time ago, with the panting professor hot on their heels, or nearly. The pilot had inspected the vehicle and checked that the controls were operational. Everything seemed okay. And the professor, by no means unawkwardly, guarded the prisoner, pointing his own weapon at him. Whether he knew how to use it was anybody's guess.

For their part, everything was ready. But the others weren't showing up.

The pilot decided to go see why. Given the professor's obvious misgivings, though, he stopped first to strap the prisoner—whose hands were tied behind his back—to the seat. Grateful, the professor felt he could relax. He even tried to make small talk with the mercenary.

Going back down the corridor, the pilot arrived behind Kathia just as Helias turned the corner and stopped short, shocked. Then Kathia sprang to one side, firing a shot as she lept. Helias threw himself to the other side, and the shot landed in Skullet's arm, catching him as he too rounded the corner. Before Skullet's weapon had touched the ground, Kathia, still in mid-air, had swiveled around and was now aiming her gun at the pilot.

"What's going on here?" cried the pilot, reaching for the pistol.

"Keep out of this—and don't touch that weapon!" barked Kathia, her voice unrecognizable.

"I demand an explanation!" he bellowed, his glance ricocheting back and forth between Kathia and Skullet.

"You're no position to demand anything. I'm taking command of the expedition, and you'll abide by my orders." said Kathia as she rose.

"By what right?"

"By right of the fact that I've got a pistol pointed right at your briefs. It's not pleasant, you know?"

Skullet, his arm aching, murmured "She's crazy….". But a fiery look from Kathia shut him up instantly.

Helias, still up against the wall where he had fallen, was, obviously, stunned and speechless.

M. Villata, *The Dark Arrow of Time*, Science and Fiction,
https://doi.org/10.1007/978-3-319-67486-5_10

Rapidly shifting her aim between the men, Kathia backed up as far as the turn in the corridor.

"Where's Junas?" she asked, addressing Skullet.

"Dunno.... He was behind me...."

"You're lying! You shot him.... In the legs. We'll never be able to drag him along behind us. Come on! Let's go. You go ahead. I've got you in my sights."

As they walked into the ship, Kathia turned to the professor, who was staring openmouthed at the scene presented by the new arrivals.

"Release the prisoner, we don't need him anymore."

The professor did so and the man moved off to the exit, through the airlock and out the armored outer door, which Kathia closed behind him.

While she was thus engaged, she turned her back for a moment on Skullet, who took advantage of the apparent distraction to draw a small weapon from a concealed holster with his uninjured arm. Uninjured, but not for long. It was immediately put out of action by another sudden shot, again from Kathia's pistol, its muzzle barely protruding past her armpit as she fired without even aiming, and still half-turned toward the door. Right afterwards, she fell to the floor with a groan, hit by a scatter beam from the pilot's weapon. She had barely time to fire off a second shot, but only struck Skullet, who went crashing against the wall.

They had 'passed' the loop. Cocooned in his helmet, Helias had heard his mother's voice giving the 'all clear' for the passage. And, even without seeing them, he knew that 'the other themselves' had passed in back-know at the same time.

Skullet was behind him, passed out. Kathia alongside, also unconscious. The professor behind Kathia. The pilot was where he belonged.

They were flying over Alkenia. After the firefight between Kathia and the others, an icy silence had descended on the ship. No one had dared speak as they heaved the bodies onto the seats and tied them there. The most urgent thing, anyway, was to cross the loop and return to Alkenia.

Now Helias and the professor were looking at each other through their visors, mutually wondering what to do next. The pilot left the cockpit, first shifting to autopilot.

"What are you going to do!?" exclaimed Helias when he saw the pilot bend over Kathia.

"Nothing special. I just have to check something...."

So saying, he began to rifle Kathia's pockets. When he got to the breast pockets, Helias could barely contain himself.

"Here they are! I knew it...."

"Here're what?" snarled Helias.

"I hardly think I owe you an explanation.... In any case, here are the two diskettes. The proof that our lovely friend wanted to make fools of us all. She had kept the compiler for herself and pretended to hand the second diskette over to Junas, but she clearly had him give it back to her on the quiet. Then, for some reason, maybe because she 'read' that somebody suspected something, or simply to get rid of excess baggage, she thought best to put them out of action before getting to the ship. She probably meant to bump off all three of you. Fortunately I arrived in time to stop her. She couldn't snuff me yet, as the pilot she still needed me. It's likely that she would have arrived at the ship alone, with some story about how you had all been captured, or worse, and later, once at our destination, she would have got rid of me and the professor too. She betrayed us all. She was clearly on the other side. Maybe we should prevent her from doing any more damage…."

And the pilot reached for his pistol.

"Stop!" ordered Helias, drawing his own.

"What are you doing? Are you on the other side too?"

"Certainly not. But I think you're rushing things a bit. After all, what you're saying is just speculation. And anyway, I'd like to hear her version before jumping to conclusions, if you don't mind."

"As you wish. I've got to get back to the helm. For me, it's enough that you tie her up, or keep watch over her yourselves. I don't want to have to think she's free to move, behind my back, when she wakes up."

The professor slowly shook his head inside his helmet, looking at Helias who was facing him. He didn't believe the pilot's version either. He raised his hand and placed his fingers at the base of the helmet. Then he pressed two buttons, in succession, making what he was doing clear. Last, he pantomimed pressing another two, motioning that this was what Helias should do. Following his directions, Helias pushed the two buttons on his own helmet. Then he heard the professor's voice.

"Good. Now we're partially isolated. In the sense that our helmets continue to receive everything, but the others can't listen to what we're saying."

"What do you think about all this, professor?"

"I don't believe the pilot's version, even if I have to admit that I'm very puzzled by Kathia's behavior."

"I agree. The pilot's, though, is suspect too."

"About the second diskette, you mean? What is it?"

"Exactly. It's a program that by generating a password makes it possible to access the compiler. Just as you had no way of knowing it existed, neither should he have known, since neither of you were there when it was handed over. And he couldn't have known it was given to Junas, unless somebody kept him informed about what was going on."

The professor turned toward Skullet.

"Very likely." confirmed Helias. "They must have been in contact by cell."

"And so now we have at least two suspects."

"Maybe Kathia had found out about their plans and tried to scupper them…. But there are some things that just don't make sense…. Why, for example, did she try to shoot me, with that crazed look on her face?…"

"What exactly happened? Tell me about it."

Helias tried to quickly fill the professor in.

"Maybe I've understood what happened. Correct me if I've got something wrong. Let's say Kathia 'read' something. Spitzer's attack on Junas, for example, or

Spitzer's intention to attack her. So she decides to 'head him off at the pass', or in other words around the corner. From what you tell me, she must have calculated everything down to the smallest details. I know her, she's capable of it. That 'crazed look' you mention is just her extreme concentration. First she drew ahead of you, to have time to prepare her ambush, not for you, but for Spitzer behind you. Even if you hadn't thrown yourself to the ground, she would have hit Spitzer anyway, swerving to the side."

"That way it makes sense, at least some of it. But why then did my father warn me about her?"

"Why don't you ask her?"

"What? She's still out cold."

"Go on. Take her hand. You don't know your own powers yet?"

Helias took Kathia's hand, and almost immediately she opened her eyes. The professor fiddled about with the helmets to tune her in too.

"How's it going?"

"Ugh, I'm a wreck. But I see our peerless healer has been at work. Thanks."

"Don't mention it."

"What happened?"

Kathia was brought up to speed, without omitting the accusations against her.

"Yeah, the dapper little blonde…. He was the first I suspected, he's not very good at covering up his thoughts…."

"Or maybe you're the one who's good at 'scanning'." said the professor.

"The other one was a harder nut to crack. He only let down his guard a couple of times, in the lab. The first when he was talking on the mercenary's cell, he 'wandered' for a second."

"Ah! That's why you were frowning…." broke in Helias.

"The second time was when he said the mercenaries were animals for what they had done to your father. But he shouldn't have known anything about it, since he had just come in from the corridor."

"So you already suspected them both."

"Yes. But I didn't know when they would have made their move. I did a bit of a shell game with the diskette, and Junas lost. I'm sorry about that. I used him as bait, in a way. But I had no choice, given what was at stake. The attack on him set the alarm bells ringing. I'm also sorry I frightened you, Helias. But I didn't have any choice there either. And now I'd say we should get the diskettes back, maybe before our friend back here wakes up."

"First I wanted to ask you about my father's strange reaction, when he saw you."

"I have no idea. I already told you. You're not convinced, are you?"

"It wouldn't be like my father to mix two people up. He's always had a good memory for faces…."

"Who knows? Maybe his sight was blurred…."

"I don't think so. He knew what he was saying."

"I think I might see what the explanation is. But it's a long story, and we don't have time now…."

"What are you up to back there? Don't you think all this silence is suspicious?" interrupted the pilot's voice. "I wanted to inform you that there's a ship approaching. It's been following us for a while."

"You two answer. I'll pretend to be unconscious still."

The professor pressed the buttons, as did Helias.

"What's going on?"

"I don't know. They haven't hailed us yet. But they will soon, judging from the approach speed."

In fact.

"Patrol ship here. Identify yourselves and slow down."

"Shuttle on special mission. Sorry, but our mission does not allow us to slow down or communicate with you."

"Give us your operation code and acting commander's name."

"Sorry, we are not authorized to give you that information."

"What agency do you belong to?"

"Sorry, but we must interrupt this conversation. We can only tell you that you should have received notification of our presence, with the order to let us pass without interference. Please check. Over and out."

"You're quite good at telling stories, I see." said the professor.

"It was the only way to gain us a little time. We can't outrun them, they're faster and well armed. The only way is a rapid change of course, right after activating the cloaking devices."

"What's the point of changing course, if cloaking makes us invisible?" asked Helias.

"The cloaking isn't perfect, not at all frequencies. They can find us anyway, if they know where to look. So we can only try to veer suddenly in another direction."

"Why didn't we 'shield' ourselves from the outset?"

"You'll understand in a moment. Go!"

"Get ready to swelter, my boy." said the professor.

It was like going into a furnace. Then the fans went on, making the heat a little more bearable. The shuttle had turned abruptly. And now it was slowing down.

"What's happening?"

"Nothing special. We've just been captured, is all. We didn't make it."

"Patrol ship. Deactivate the shields. You are under arrest. In a moment we will come alongside and tow you. Don't try to find an escape route. You have none."

"I formally protest and am officially recording this conversation. You are willfully interfering with a special mission, and I hereby inform you that you will suffer the consequences."

"Note duly taken. As for your mission, we have made inquiries and the records show nothing of the sort. Prepare for docking."

"One moment!" broke in Kathia. "We have two casualties on board. On a Code Delta basis, we ask that you transfer them on board your ship so that your medical officer can proceed with emergency care."

"Granted. Prepare for docking and transfer. Over and out."

The pilot had left his newly useless cockpit and approached the trio. "And so, if I'm not mistaken, your sudden recovery means we need another casualty, right? Were you thinking of me, by any chance?" he said, drawing his gun.

"Don't be a fool. It's three against one here. Give me the diskettes."

"Three? You're unarmed. As for the other two…." And he scoffed at them with a mocking smile.

Helias, half hidden by the seat in front of him, was trying to pick up his weapon without being noticed. But the expression on his face had changed as a result. And Kathia understood that he would never manage. In fact, the pilot had realized what he was doing too, and now turned his entire attention to him. As well as his pistol.

Kathia took immediate advantage, catching the pilot with an explosive kick to the nether regions, and at the same time diving on top of Helias to retrieve his weapon.

But the pilot had already fired off a shot, toward her as she sprang from one seat to the other. And so the shot hit slightly off target, in Kathia's left shoulder, which started to gush blood. This time the weapon had been loaded to kill. Kathia, wailing with rage and pain, jumped to her feet and, supporting herself against the seats, landed a kick in the middle of the pilot's chest. With a whimper, he crashed to the floor, banging his head.

"And now three of us are casualties, you son of a bitch." Kathia just managed to get the words out, in a voice that was not her own, before sinking exhausted into her seat.

Kathia had hurriedly stanched her wound, summoned her last strength, and headed toward the control console, stepping over the body of the pilot, who was already coming to his senses.

She entered a few commands on the screen, then summoning Helias.

"Stay here and don't move, whatever happens. At a nod from me, press this blue button. After a few moments it will start flashing. The flashing should last around thirty seconds, and during that time you can push this orange button if I say 'Now'. Otherwise the time will run out and we'll have to repeat the operation. In any case, after the orange button, get strapped down as soon as you can."

Helias was about to ask for an explanation, but the patrol ship had docked and two uniformed men were already appearing in the airlock door.

"Hands up, all of you, and no false moves. You're under arrest. Any attempt to resist on your part will authorize us to shoot."

"Easy. As you can see, there's not much that's able to resist here." said Kathia as she put up her hands, imitated by Helias and by the professor. "I ask you only, as the first thing you do, to take Professor Borodine here directly to your commander. The professor can explain the delicate situation we're into him privately. Then the unconscious passengers should have precedence in being transferred…. Lastly, you can treat my own wound, and take us all down to the base…."

By now, Kathia was clearly having trouble speaking, and couldn't stand without leaning on something.

Two more patrolmen came in, to help with the transfer. The professor had gone first, after exchanging a few glances and a few comments in an undertone with Kathia, who seemed to confirm that his impressions of what she was planning were correct. Then it was Skullet's turn to leave the shuttle, still unconscious on a stretcher. The pilot, after a quick check with the radioscope, was helped to his feet and, staggering, was conducted toward the exit.

Kathia followed him slowly and at a distance. And Helias understood that she was calculating the next movements, probably also reading the patrolmen's minds. A few steps short of the doorway, she turned toward Helias, who was waiting near the console for her signal. But one of the two men guarding the door—who, holding the shuttle's passengers at gunpoint as they frisked and disarmed them before allowing them through—had his suspicions roused when he noticed Helias hanging back near the control console.

"What are you doing there? Come along!"

Kathia was forced to jump into the breach. She threw herself on the still-dazed pilot, knocking him atop the two guards, who didn't have time to react. At the same time, she shouted "Now! Right away!". The engines started, but the shuttle was still under the patrol ship's control. Kathia snatched up a weapon that had fallen to the ground and, with a look that quelled all protest, relieved the third remaining patrolman—the only one still on his feet—of his gun. Then, aiming the two pistols, she ordered all three to disembark without delay.

While this was going on, Helias had pressed the two buttons in succession, and was now standing agog as he watched Kathia in action.

Seeing him gawping at her, she shouted "What are you waiting for? Get strapped down!".

And then it hit him: Helias finally understood. He understood that Kathia was about to go through that door. And that maybe he would never see her again, ever. Because only she, from the patrol ship, could free the shuttle.

And in his mind's eye he saw his mother again, going through that other door.

And the look Kathia gave him was not unlike his mother's. A look that said 'I'd give anything to stay here, with you.... But I can't.... Because it would mean losing everything. Because I have no choice. Forgive me....'

And so Helias stole this one last look. Then he strapped himself to his seat.

He clenched his teeth and shut his eyes, his throat tightening. But he couldn't prevent a tear from filtering past his closed lashes. A tear that, as the ship accelerated, was pushed back in a thin bright line across his temple and into his hair.

And so Helias was alone. With the befuddled pilot. And running on autopilot, as set up by Kathia. He wouldn't even have known where to start. He could only hope that the pilot down there, still flattened under the weight of acceleration, would come to quickly. In case corrective manoeuvres were needed. And to get somewhere, and not just escape.

How they had miraculously escaped from the patrol was something he didn't understand. Doubtless it was Kathia's doing, and perhaps the professor's too.

Helias didn't even know whether the Alkenian patrolmen were earthlings or Thaymites. In either case, they had too much to hide from them, whoever they were, and absolutely couldn't risk letting the diskettes fall into the wrong hands.

So Kathia had chosen the lesser of two evils, enabling him to get away. Him and the pilot, who had the diskettes and, what's more, was indispensable for flying the shuttle and thus bring Helias and the diskettes to safety. But Helias had to be able to deal with the situation. And to get the better of the pilot. Kathia, then, had had faith in him. Or maybe she had just decided that she had to have faith. As the last resort. As the only way to accomplish the mission.

Helias released his harness, rose and went to level his gun at the pilot, who just in that moment was getting to his feet.

"Right, we're alone. It's between you and me now. Hand over the diskettes."

"Diskettes? What diskettes? Ah, the diskettes.... Why? And what're you going to do if I don't?"

"I've got a gun, and you're unarmed...."

"Oh, yeah? So you'd shoot me.... And then what? Who'd fly the ship?"

"Me."

"You? You're bluffing pathetically. Get out of the way, I've got things to do."

And he went to the control console, pushing buttons here and there.

"One of our engines is out. What happened?"

"We were blocked by the patrol ship."

"That I remember. What happened afterwards?"

"Kathia and the others went to the other ship. I started the engines, or at least I think I did.... After Kathia had fiddled with the controls. Then the hatch closed, and almost immediately the shuttle rocketed off like a torpedo...."

"Hmmm, I think I see. In addition to freeing us from the patrol ship's control, Kathia must have managed to prevent them—at least temporarily—from following us or capturing us again. But we can expect another attack, sooner or later. So let's start cloaking. I'd advise you to take off your coveralls."

That suffocating heat again. Helias started to take off his coveralls. But he had to lower his weapon. And once he realized his mistake it was too late.

When he came to, he was stretched full length on the floor, disarmed, legs paralyzed. And, if that weren't enough, half undressed, soaked in sweat, and with a splitting headache.

"Oh, welcome back!" smirked the pilot, bare-chested at the controls. "I was beginning to think you'd never wake up. As you can see, I went easy on you, I could have done a lot more damage. But it wasn't worth it. And you might come in useful."

"Where are we? How much time has gone by?"

"Several hours. First we quickly reached the cone of shadow, and so I was able to turn off the cloaking devices and return to a normal temperature, turning them back on only every once in a while to prevent a systematic search from finding us."

"But where are we?"

"Not far from the Scientific Station. A bit before our arrival from Thaýma. Remember?"

Actually, Helias was still pretty woozy, and was having a hard time connecting. Eventually he said "Yes, I remember…. The black object, the counterpart we couldn't see….".

"Right. That's us."

Sensation was returning to Helias's legs, which had begun to tingle. He stealthily reached an arm under the seat, to retrieve the weapon the pilot had dropped when Kathia felled him with a kick, and which everyone seemed to have forgotten.

The pilot was concentrating on the 'passage' through the loop. Which Helias, still stretched out on the floor, practically didn't notice. Then he felt a sharp turn. And understood the pilot's intentions. It was a manoeuvre he hadn't known, yet. He asked for confirmation.

"Where are we? What do you intend to do?"

"We're on Thaýma. I'm drawing alongside the cargo ship. We'll pass together with her. With a bit of luck."

They passed.

By now Helias had fully recovered the use of his legs.

"Now where are we headed?"

"To the capital."

"Do you mind dropping me off at the Kusmiri Center first?"

"Are you joking?"

"Hardly. Seeing that I've got your pistol pointed at your head." said Helias as he rose.

"Ah! That nonsense again?… How do you think you're going to land, if you shoot me?"

"Bah! I'd say at this point I don't have a lot to lose, don't you think? Either we crash now, or sooner or later you ice me. Don't have a lot of choice, do I? For you, on the other hand, it's a question of living or dying…."

"So let's make a deal."

"Let's hear it…."

"I let you out wherever you want, and then you let me go off on my own. With the diskettes, naturally."

"I don't see why I should agree to that…. Since I'm holding all the aces now…."

"Careful…. You already thought you were once…. And you saw how that ended…. I've got a thousand ways of disarming you. A sudden turn, for example…."

And he matched his words by sharply canting the ship for a second, which nearly sent Helias reeling.

"Steady!"

"Take it easy! It was just an example. You'd better take me up on my offer. Or it will be so much the worse for you…."

At that point a red lamp started to flash on the console screen.

"Shit! We've got problems with the other engine too. The patrol ship probably damaged both. Just that one of them hung in there until now…."

"What's this? Another of your tricks?"

"No, unfortunately. And it's serious. We've got to land right away."

"No funny business…."

"I've got a lot more, and a lot worse, to worry about at the moment."

And suddenly, the ship went into a nose dive, almost in free fall, throwing Helias's legs out from under him.

Leaving the controls, the pilot pounced. A shot rang out. And a moment later the pilot's body was floating in midair, as the shuttle continued to plummet.

Helias struggled to the console, latched his harness, and tried to concentrate on the screen. At the end, he decided to chance everything on one control that seemed his best bet.

The shuttle began to lose altitude less quickly. Gravity returned, and something that bore at least a passing resemblance to stable, horizontal flight.

But by now they had lost too much altitude, and the shuttle's course, still angling down, barely cleared several snow-capped peaks, and was heading into a wooded valley.

Helias managed only to pull away from the mountainside at the last minute, avoiding a crash by inches.

The shuttle began to snap off the tops of the trees. And larger and larger trunks, the closer it got to ground. Then, its impact force exhausted, the crippled vehicle caromed off a spinney and careened across a clearing, flipping over and over until it plowed to a halt.

Silence fell. In the pallid light of Alkenia's red sun, just clearing the mountains. In the middle of nowhere, far from the capital and the Kusmiri Center.

Chapter 11
Nothing Moved in the Silent Valley

Nothing moved in the silent valley. Nothing but the leaves, fluttering in the tiniest of breezes. If it weren't for the freshly snapped tree trunks and the splintered branches scattered round, no one would have suspected that the battered fuselage lay, equally silent, even more motionless, down there in the clearing. Lifeless.

And then, inside the fuselage, something began to move. A groan was heard. Someone was regaining consciousness, hanging head down. The click of a harness unlatched. And a body plopped to the floor, or rather, the shuttle's overturned ceiling. Almost where another body lay dead.

The live body detonated the emergency door release and dragged itself out, slowly. It crawled a dozen yards from the wreck, stopped, and keeled over onto its side.

Helias woke up a few hours later, the sun high in the sky. The first thing he felt was a sharp pain above his right armpit, between shoulder and arm. He touched it. The shoulder was bare. Then he remembered he had been trying to undress, before the pilot attacked him. He was losing blood. And there was a long cut, back where he couldn't see it, and it burned.

He got to his feet and went back into the shuttle. To get the backpack with Kathia's medical supplies. He tried to disinfect and treat the wound, but it wasn't easy reaching back with one hand.

He had various other scratches and bruises, but nothing serious. He moved away from the ship again, still staggering, the top of the coveralls flapping around him, since he couldn't slip it back on because of his injured arm. He sat on a flat rock a little further down and started to look around, wondering where he had ended up. He supposed that none of the instruments on board were still working. He wouldn't have known what to do with them anyway. He tried the signaler on his belt. At least that was working, or so it seemed. He activated it and, even if he wasn't superstitious, crossed his fingers. He hadn't the faintest idea what its range might be, or how far away he was from a receiver, any receiver. Or what frequency it transmitted on. All he could do was hope. That his feeble message, something like

M. Villata, *The Dark Arrow of Time*, Science and Fiction,
https://doi.org/10.1007/978-3-319-67486-5_11

'I am Helias Kadler. I am lost and in difficulty. Please rescue me.', would reach somebody, anybody. He also tried the cell but, obviously, there was no signal.

Then he looked at the mountains around him. Maybe, from up there, his signal would go farther. It's not as if he had anything better to do. Than collect his strength, find some food, and something else, in the shuttle, and get going. Slowly.

He was almost down to the valley floor and, from where he was standing, there was a fairly good view of the opposite slope, which didn't seem quite as steep or as rugged as the one he was on, or at least as far up as he could see, which wasn't much. And halfway up the opposite side, it looked as if the slope flattened out into a plateau, or at least a basin or high valley, which would certainly make for easier going.

Fast racks of clouds kept scudding across the sun. It was colder now, and the wind had also come up, rattling the leafy branches. Helias donned the parka he had found in the shuttle and slogged across the broad shallow stream that ran through the bottom of the valley.

He started climbing. First along a wooded slope. Then he picked his way along the scar left by a landslide. And then he crossed the rubble, following the contour of the ridge it had fallen from. This brought him to a gentler slope, where the trees started to thin out.

Now the clouds were massing, thick and threatening. And, turning around, he saw that they had covered the other side of the valley, hiding the peaks and upper reaches from sight.

He was tired. And the cut burned. More than ever. He ran two fingers under the bandage, and could feel that it was swollen and aching. And the weather kept worsening. It was cold, and the sky was grimmer by the minute. The first drops came, and the first thunderclaps. He decided to look for shelter: weather apart, he could barely keep his feet by now, and he could feel a fever coming on.

He huddled under a boulder that projected a bit more than the others, as a heavy rain began to fall. And darkness fell too, torn apart by lightning flashes in rapid staccato. Suddenly, Helias felt alone, more alone than ever. Alone in the dark, with the cold gnawing at his bones and rattling his teeth. Alone with the constant crash of the rain, constant except when overwhelmed by the thunderbursts whose echo faded away among the farther mountains.

When the storm went raging off elsewhere, and the shredded remnants of the last laggard cloud paled in Nasymil's icy light, then, and only then, Helias, sapped by fever and exertion, slept.

Shortly afterwards, the sky had cleared completely, and the starlight cast a shadow on the rock that sheltered Helias's slumber. The shadow of a hooded form. Which bent over, sedated him, and searched his pockets. Finding what it sought, it slipped his parka half off of him, eliciting a whimper of pain. It unbandaged the wound and placed a hand over it, for almost a minute. It sprayed and cleaned the gash, covered it with a damp dressing, topped by an adhesive membrane. It pulled the parka back up and paused for a few moments to contemplate that wan and febrile profile. Then, noiselessly as it had come, it left.

Some time went by. Helias wouldn't have been able to say how long he had slept. He was woken by a low rumble, from farther up. The rumble of an engine.

Then he heard voices. And saw lights approaching. Deliverance. He called out.

"Hey! I'm over here. Kadler. Helias Kadler."

The lights shone on him, blindingly. Two rescuers. The nearest leaned over him and dimmed the searchlamp. Now Helias could see too. And, with the other searchlamp's light reflected by the rock, he recognized the first rescuer. The blond hair, the smiling face were Kathia's. But his sight was already clouding over. And Helias sank back into sleep. Serene, relaxed, a smile on his lips.

Helias woke in a room with two beds, separated by a screen. Some sort of infirmary, or a hospital ward. The other bed was empty.

He felt well enough, but a bit disoriented. He touched his shoulder. No pain, and just a light dressing. He got up and went to the window. He recognized the inner courtyard right away. He was in the Kusmiri Center. Not far from his old room. And a whole series of questions rose to his mind.

While he was still at the window, he heard someone come in. He turned and saw Kathia. He went to her, embraced her. She returned the embrace.

"Welcome back." she whispered. Then she looked him in the face and kissed him. And then looked him in the eyes again, caressing his face.

"I felt that you were awake and I came right away. I also hear all the questions mulling around in your mind. And I'll try to answer them."

Helias looked at Kathia, as if he were seeing her for the first time.

"Do you find me changed? Well, since we parted, there on the shuttle, a little over a day has gone by for you, but for me it's been more than three years. I've been waiting for you for three years. And I couldn't wait to see you again. What have I been doing in these three years? Much of the time I spent in prison. The patrolmen didn't much like the little trick I played on them up there. In fact, I was a bit rude, and I infringed I don't know how many articles of the code. And so I was charged with all those counts and sentenced to three years, but they let me out early for good conduct. The professor got off more lightly, eight months on a suspended sentence. I don't know what became of Spitzer. It's as if he had evaporated. Who were the patrolmen and who sentenced me? Thaymites, all of them. No, the earthlings had nothing to do with it.... No, I couldn't ask for help. And help from who, anyway? The counselor? I would have put him in quite a spot, don't you think? And anyway, I clearly didn't ask him, otherwise he would have talked to me about it, before our departure. Or at least I think. Where have I been? On Thaýma. The professor too. With the false IDs we had brought with us, just in case something like this happened."

"What day is it today? The day after our departure?"

"Right. You—or rather, we—were away from the Center for slightly over one night. Except for the professor, who returned yesterday evening, a little after the departure. And who's already filled in the counselor."

"So we came back more or less together...."

"I'd say exactly together. We were on the Thaymite cargo ship. Expressly for your return. Even though we couldn't be certain, it was pretty likely that you were about to arrive, sooner or later. Given that three years ago no trace of you could be found on Alkenia, the most probable thing was that you had passed on Thaýma, at the same time, but in the opposite direction, as our 'old' back-now passage. That would also explain the business of the 'black object'. As it happened, we chanced it, and got lucky, it would seem. We also knew that your engines were presumably damaged. As soon as we had passed on Alkenia with the cargo ship, we had a shuttle come pick us up from here, and we went looking for you, first coming here to drop off the professor. And so at a certain point we located you. And here we all are. Incidentally, where did you get the material you used to treat your wound?"

"From your backpack on the shuttle."

"Ah!"

"What's the matter? You're not convinced?"

"No, nothing. Just asking."

"And Mattheus?"

"Didn't I tell you? He was there too last night, looking for you."

"Where have you put my coveralls? Did you get the diskettes from them?"

Kathia looked at Helias keenly.

"No. No diskettes.... Maybe you lost them.... I know you took them from the pilot, I read it in your mind while you sleeping."

"That can't be. They were in a closed pocket...."

"A side pocket?"

"Yes."

"We found both of them open...."

"Who was with you, in addition to Mattheus?"

"Two of our own men, both trustworthy."

"Like the pilot and Spitzer?"

"No. I checked them out myself."

"They might be sharper than you suspect."

"I don't think so."

"And Mattheus? How good is he at masking his thoughts?"

"The best I know. What is it? Don't you trust him?"

"I don't know who to trust anymore...."

"Not even me?"

"You? That's another thing."

"And what kind of thing might that be?"

"You know perfectly well, since you read my thoughts."

"Actually, I'd say that your thoughts on the subject are quite muddled at the moment."

"I suppose it's because you seem so changed...."

"Over two and a half years in prison can change a person a lot...."

"How much?"

"No. Not the way you think. Not in questions of the heart...."

And Kathia kissed Helias. And caressed him. Unhurriedly.

"Maybe it's better to stop here." Kathia said at last.

"Yes, also because somebody might come in."

"What were you saying about Mattheus?"

"Nothing specific. Just that there must be something fishy, somewhere. First the two imposters, Spitzer and the pilot. Then the disappearing diskettes.... Who else would have been able to search my coveralls, before you?"

"There were only the four of us. I searched you slightly after you were embarked. As soon as I 'read' that you had the diskettes."

"Who of the others would have been able to do it first, without you seeing him?"

"I don't think I can rule out anybody.... I wasn't able to keep my eye on you all the time, for various reasons."

"I'd say it calls for an investigation."

"Without ruling anybody out, you think...."

"You know them all better than I do...."

"I've known Mattheus forever. I'd trust him implicitly. It can't be.... Unless...."

"Unless what?"

"No, nothing. It can't be.... Nothing, never mind. Just a crazy idea...."

"Tell me anyway."

"There's another odd thing. That you don't know about yet."

"Being?"

"Remember that before leaving, you told me that you had left the red diskette for Mattheus? I remember you were sincere...."

"And so?"

"And so Mattheus maintains that he didn't find anything in the place you indicated."

"And you don't find that suspicious?"

"I don't know. You both seem sincere. Maybe you two misunderstood each other...."

"I really don't think so. Just as I don't think somebody could have preceded him."

"So as you see it, he took it, with the intention of not giving it back...."

"Or to give it to someone else.... Would he be capable of hiding something like that from you?"

"Maybe he might...."

"We have to find out. Where is he now?"

"I don't know. Probably over there, inside the mountain."

"Let's go...."

"Wait a minute. First let's check that there wasn't a misunderstanding, and that the diskette really isn't there."

"Right, good idea. Let's split up, and one of us check the hiding place and the other check Mattheus, before it's too late. If it isn't already."

"But we can't approach him cold like that. He'd see right away that we suspect him. If push comes to shove, it's better that both of us be there, maybe with reinforcements."

"Okay, so you start by checking the hiding place. I've got to check something else, so I need the key to my room…."

"It's over there. I'll go get it."

Helias went back to the window, looking strangely exhausted.

Time passed, and Kathia still hadn't come back. It was taking her longer than he had expected. Helias left the room to look for her. He passed a door that had been left ajar and heard her voice. She was talking on the cell, in an undertone. He couldn't make out what she was saying. Helias knocked.

"Come in…. I'll be right there." she said to Helias. Then she said something in another language, incomprehensible, but which probably meant "Wait a second." addressing whoever she was talking to on the cell, but still looking at Helias.

"Here's the key. If you'll wait a moment, I'll come with you."

"It doesn't matter. Just give me a call. Is your code still the same?"

"Yes. Here, catch."

"Talk to you in a bit. In the meantime, check the hiding place…."

"Helias…. Where is the hiding place?"

"What? You don't know?…"

"Well, no. I know about it, but not where it is, exactly."

"Sure…. I'll explain later. Call me when you've finished."

"Okay. Go to the left. At the end of the hall, turn right. You'll end up in the inner courtyard. After that you know the way."

"Okay. Talk to you later."

Helias went out, again leaving the door slightly ajar, and headed off to the left. Then he turned on his heel and went in the opposite direction, making no noise. He passed a corridor branching off, and then another. He took a left and was about to go out into the courtyard. But stopped in his tracks.

He had recognized an unmistakable figure in the half-light of a nearby foyer. It was Geremy Stuerz. Who was talking on his cell.

Helias went back the way he had come and followed Kathia's instructions. Passing the unclosed door, he heard her voice again.

Back in his room, he turned out the contents of a drawer and found the 'bandaid' cell.

"Saileh…. Professor, can you hear me?"

No answer. And the other cell rang.

"Here I am. Sorry about the delay."

"Shall I give you directions to the hiding place?"

"Wouldn't it be better to go together? Or for you to go yourself?"

"Together we'd attract too much notice. Don't forget that they might be staking us out. In fact, I'd say it would be better if this conversation lasted as little as possible. So I'll tell you only what's strictly necessary. At intervals. It's better for me not to go. You're more likely to notice if somebody's watching you or following you."

"Okay. Tell me."

Helias fell silent for a few moments.

"Hello?"

"Yes, I'm here. This is the starting point, then I'll give you further directions. 'Where what sees is transformed you will find it.'"

Silence. For a few seconds.

"Right…. But a lot of time has gone by, for me. I don't remember too well. Be more specific, please."

"Leave the building, toward the lake. Go right and follow the shore as far as the bridge. Then call me again."

"Right. Over and out."

She hadn't objected!… Nor, over the cell, could she realize that Helias was deceiving her. That's exactly why he had chosen to communicate that way.

Helias watched from the window. He still couldn't believe it…. Kathia really was walking along the lakeshore toward the bridge, in the direction opposite to 'their' rock. Unhesitatingly.

"Saileh…. Professor, can you hear me?"

"Yes, I hear you. Welcome back, young man! The 'bandaid' automatically sent a message to my cell, and here I am. How are you doing? I heard you had a pretty close call and were injured."

"Nothing serious. I need to talk to you urgently."

"Let's hear it."

"What do you think about Kathia? I mean, doesn't she seem strange to you, changed?"

"Yes. She's changed a lot since leaving prison. Which can be fairly normal. The strange thing is that she changed suddenly. When she got out, not while she was a prisoner, when I went to visit her occasionally."

"Could they have brainwashed her, or something of the sort?"

"Honestly, I wouldn't know. Why do you ask?"

"Nothing I can pinpoint. But everything in general. She's like someone trying very hard to be herself, or rather, to interpret an image of herself. But without really being able to."

"It wouldn't be the first time that that girl has surprised us…."

"Now it's different. She's missing something. She also seems to have lapses of memory, pretty big ones."

"Perhaps we'd better talk about this in person."

"Yes. And I also have to ask you about Mattheus."

"Is he 'strange' too?"

"Maybe worse."

"I don't know, I haven't seen him yet."

"Neither have I, but there are…."

Helias's cell had rung.

"I've got to answer the cell now. It must be Kathia. We should call or meet later."

"Fine. I'm back in my office. I'm here. Over and out."

"Yes?"

"It's me."

It was Kathia, but her voice was strange.

"Helias…. You lied to me. You owe me an explanation…."

"I'd say I do."

Helias was pleased. Kathia appeared to have recovered, at least partially, her memory of those days. Maybe he had been too suspicious. And it wouldn't have been the first time he had suspected her wrongly. They just had to talk about why she had been acting so strangely. Maybe there was a simple explanation….

"Right. Let's talk about it now."

"Yes. I'll come to your room in a minute."

"I'll be waiting for you. Over and out."

Minutes ticked by. More than enough time to return from the bridge. Helias went back to the window. No sign of Kathia on the lakeshore. Perhaps she was arriving. He went to the door. There was no one in the hall.

Other, endless, minutes passed. Then someone knocked. He looked through the peephole. It was Kathia. He opened the door.

"Hello, Helias."

Kathia had a strange look on her face. And there was someone at her left. Mattheus. Then his sight fogged over.

And the last thing he saw, a hand's breadth from his nose, was Geremy Stuerz's idiotic grin.

Chapter 12
Helias Was Stretched Out on the Floor of His Room

Helias was stretched out on the floor of his room. Half conscious. As if he were sleeping, but he could hear the voices and the questions. And his mind answered automatically, he couldn't help it.

"He really doesn't know anything about the other two diskettes." said Kathia. "But we have to get them back too. Those are our orders."

"Why isn't the other one enough?" asked Mattheus.

"Because the idea is to keep anyone else from using them."

"But this fellow doesn't know anything…."

"So we have to look elsewhere, and…."

A thundering at the door, and a voice resounded from the corridor.

"Open immediately, or we'll break the door down!"

"How many are there, do you think?" asked Geremy Stuerz.

"I feel four, I believe." answered Kathia.

"Maybe five." said Mattheus.

"What are their intentions?"

"Bellicose."

"What should we do?"

"Answer them." said Kathia. "What do you want?"

"We're the ones who are asking the questions here. Open immediately."

"Easy, easy. We're coming."

Kathia opened the door, which burst open.

"Freeze, all of you! Or you'll regret it."

Four armed men strode in, followed by the professor.

"That's enough now. Wake the boy." ordered the latter.

Kathia motioned to Mattheus, who sprayed something under Helias's nose. Which woke him almost instantly.

"What's going on? Kathia! Mattheus! Why?… What are you doing with this guy?"

"Arrest them." said the professor. "We'll interrogate them later."

And he bent over Helias.

M. Villata, *The Dark Arrow of Time*, Science and Fiction,
https://doi.org/10.1007/978-3-319-67486-5_12

"Everything all right, young man?"

"Y...yes, I'm fine. More or less. But what's going on?..." stammered Helias bewilderedly, eyes darting between the three figures standing around him and peering searchingly at Kathia, who shied from his gaze.

"I'll explain. Come with me." And, to the guards, "Take them away. Without attracting attention. Go through the cellars."

"What do you think I've been doing all this time, these three years, young man?" asked the professor from the comfort of his office armchair.

"I've looked into things, studied the situation. I had three years to the good, before taking up from where I left off. Might as well use them to try to understand what was going on, I thought."

"Fine. But first explain Kathia's and Mattheus's behavior to me, please."

"Don't rush ahead. We're getting there. But you'll understand better if we start from the beginning...."

Helias shook his head.

"You can't wait, can you? You're in a hurry to know? Well, I see your point."

The professor remained in silence for a while, stroking his wattles.

"In a certain sense, I'm about to give you good news. But only to some extent. I'm not sure they'd approve. Are you certain you want to know?"

"Absolutely. Whatever it is."

"Right. The ones you saw before, and since you got back here, are not...."

"They're not the same! Of course they're not themselves! What an idiot!... But then...."

"Yes, my boy. They're clones."

"Clones of who? They're practically the same age...."

"Clones of themselves...." murmured the professor, studying Helias from behind his lashes.

"They're copies.... Cookie-cutter twins.... They're experiments.... That's putting it brutally, but that's the way it is...."

"Even the ones I knew myself? Kathia and Mattheus too?"

"Yes."

"But who allowed a thing like that?"

"Thaýma. Many years ago."

"Okay. Tell me from the beginning now, please."

"A sensational discovery was announced. An individual, who wished to remain anonymous, carried a genetic mutation that enabled him, sporadically and under certain conditions, to 'guess' what other people were thinking. The scientist to whom the individual had voluntarily submitted his case immediately asked to be able to conduct studies in order to isolate and reproduce this genetic mutation. The request was highly controversial, and nothing more was heard about it for some time, at least publicly. It was thought that it had come to nothing, as indeed was to be expected. A few years later, however, it was back in the spotlight, but as a done deed, essentially, a study at the advanced stage that was already giving its first fruits. But I've not been able to find out anything about the developments that led up to this stage."

"You mean to say that it's not known how or when the project came to be authorized and funded?"

"Exactly. It's strange, but there's a yawning gap in the paper trail. Anyway, to make a long story short, thus were born the first 'Mattheuses' and the first 'Kathias', born from who knows what kind of experiments in genetic engineering, that went well beyond the simple reproduction of certain gene sequences."

"How long did these experiments last?"

"Don't be fooled by the apparent difference in age between Kathia and Mattheus. In reality, they were 'born' only a few years apart. It's just that the Mattheuses were affected by a slight form of premature aging. So the Kathias were 'produced', correcting the earlier error and resulting in exceptionally slow aging…."

"In other words?! How old is Kathia, then?"

"Almost forty, I believe, or slightly less."

Helias gaped at him in surprise.

"Amazing, isn't it? In fact, the one you met would be three years older by now. We're talking about Thaymite years, but the difference with either the Earth or Alkenia is minimal…."

"Would be? Why did you say 'would'? What's happened to Kathia?"

"If I knew I would have told you right away. As I said earlier, I'd also noticed the 'change' in her, though it was only a vague suspicion that was only confirmed today, by you and by what's happened. But I had already tried to investigate whether she might have disappeared and been replaced. I haven't found out anything yet, but others are working on it, with redoubled diligence, given our recent discovery."

"Be honest with me, could she…? Could they have…, have harmed her?"

Helias was feeling terrible, visibly so. He felt as if he had had to swallow, in one horrible gulp, all of the awful surprises, sorrows and griefs of an entire lifetime.

"Can I get you something? Shall we continue later?"

"I'd just like to disappear, for an instant. Or sleep and not have to think of anything for a year on end. But I'll settle for a visit to the lavatory. And a glass of water, if you don't mind."

The professor poured the water and waited for a few minutes, doodling on a sheet of paper.

"When are you going to interrogate the impostors?"

"Shortly. Do you want to be present?"

"Yes. What's become of Mattheus?"

"We'll have them tell us that too."

"In the meantime, shall we continue?"

"Yes. Where were we? Ah, yes. Kathia's age. Don't be too upset by that. In fact, biologically she's actually twenty-five or thereabouts. Just that she has more experience of life, greater maturity. Not bad, no? Both for her, and for, uhm… you…. Because you'll see her again, I can feel it. She's got nine lives, that girl. And nobody can gain from doing her any real harm."

"Are there other 'cases', apart from the Kathias and the Mattheuses?"

"Yes, so it appears. But personally I don't know much about it. For example, that Stuerz, you knew him already?"

"Yes. From college, on Earth. A slimy sort.... And incredibly obnoxious...."

"Who knows, maybe they had already put him on your tail back then. Like Kathia, I mean. But by the 'others'."

"But who are these 'others'?"

"No idea, still completely in the dark there. In that respect, my three years haven't gotten me much of anywhere."

"How did you deal with finding yourself in a past, part of whose future you already knew?"

"As best I could.... It's not a trivial problem.... Knowing, for example, that an hour and a half down the road there's another me, living in blissful ignorance of all this. You get these urges to go shake things up for him a bit. But you already know that you didn't do it. And you certainly don't want to discover that you didn't do it because you had an accident along the road. So you cut yourself off. A sort of voluntary house arrest. Far from anybody who might recognize you and ask you embarrassing questions. Just think that before coming to Alkenia, the first time, I mean, or in other words before our time trip, a student of mine told me that he had seen me in a place where I had never in fact been. At the time, I thought it must have been someone who resembled me. Later I understood."

"So, going back to our bioengineer and his team, they...."

"No, wait. I said nothing about a team. Because there was no team, as far as I know. He, a certain Professor Winkler, appears to have worked alone. Which, at the very least, is odd...."

"Without assistants?... And so without any direct control? And no one gaining the experience to be able to continue with the experiments later?"

"So it would seem. Nobody except for a very few non-scientific auxiliaries. Strange, to say the least, isn't it?"

"Where is this fellow now?"

"Dead and buried. For more than fifteen years."

"And the research?"

"Finished along with him. Or rather, even before his death. There was a crackdown and the experiments were no longer authorized, for obvious ethical reasons."

"But in the meantime, a certain number of individuals, children of that lab and all to some extent strange, strolled the streets of Thaýma as free citizens. Is that right?"

"More or less. Free, yes, but treated with wariness. Even if they are quite useful for certain kinds of work, as you can imagine."

"How many of them do you suppose there are?"

"I don't know. A few dozen, maybe a hundred or so. The documentation about it isn't accessible, as far as I know. Not accessible to me, anyway. But I've already briefed the counselor, who'll find out. I've already told him everything yesterday evening. And a short time ago I filled him in on the latest developments. He's doing what he can for Kathia too. Meanwhile, since yesterday evening we've been organizing a group of people who will enable us to defend ourselves from these offensives, which are something new, starting with the pilot and Spitzer, down to the false Mattheus and Kathia, with that other one, who were arrested thanks to the guards in this group."

The professor looked at the time.

"It's nearly time for the questioning. Shall we go?"

"One last question. So, contrary to what I thought at the beginning, the 'normal' Thaymite population is just like Earth's, with no strange powers. Only a tiny percentage, lab children, have the strange things like Kathia's eyes and all the rest…."

"No, the eyes are another story. They're the common heritage of all Thaymites. Though Kathia's, and those of her 'brethren', have a further peculiarity…."

"Ah! However, what I want to say is: why didn't Kathia, and you too, all of you, tell me the truth right away? Why did Kathia always pretend she was an ordinary Thaymite woman when she was with me? Maybe she never said so in so many words, but that's what she always led me to believe…."

The professor was silent for a while, fingering his greasy nose.

"Because she was ashamed…. Because it was her dream…. Because she loved you. And she was ashamed, my boy, to tell you that she was only a fake woman…. Maybe she would have liked to be able to introduce her parents to you…. And she didn't want to pull a test tube out of her pocket, so to speak…. It was fun, when she started…. Because it was her dream…. But then she no longer had the courage to tell you the truth…. Which, among other things, is that she can never have children, because she's prohibited from doing so…. But perhaps that was exactly what, for the first time, she felt she wanted from a man, ever since she 'spied' on you in the college dorm…. And maybe, for the first time, she felt all the weight and sadness of her unusual fate."

Once again, Helias Kadler pictured Kathia's face, looking at him as she disappeared beyond the airlock door.

And that evening he sobbed aloud as he stood clutching the sides of the sink in the professor's bathroom.

Chapter 13
Are They High Enough Yet?

"Are they high enough yet?" asked the professor.

"High as a kite." replied one of the two men, as he sorted the vials.

All three were laid out motionless, each on a gurney, in a minimally furnished room.

A woman came in, followed by the counselor and another, rather self-important, man who was introduced as the counselor's secretary.

The counselor walked up to Helias.

"Hello, Helias. How are you doing?"

"Not bad, except...."

"Yes, I know how you must feel. Cheer up, we'll come out on the other end of this...." and he patted his arm.

Helias nodded, without speaking.

"Who shall we start with?" asked the woman, young and quite good-looking, as she regarded Helias with an intrigued smile.

"We'll start with her." answered the professor.

"She's 'forty' too?" whispered Helias in the professor's ear.

The professor nodded. "The 'brunette version', I call her. But it's no use whispering, she can 'read' us anyway."

And the 'girl' turned, and smiled.

"Better put the other two to sleep, so they don't hear." she said to one of the two attendants, who started shaking a vial.

"Let's record everything, please." directed the counselor. And the secretary produced a small camera.

"Can she lie?" asked Helias.

"Lie, no, that's almost impossible. However, she may have been conditioned not to answer certain questions. We'll know in a minute." answered the girl. "My name is Athika, since you're wondering. And you're Helias. Pleased to meet you." And she smiled again.

"Everything ready? Please, feel free to begin."

M. Villata, *The Dark Arrow of Time*, Science and Fiction,
https://doi.org/10.1007/978-3-319-67486-5_13

Helias and the professor looked at each other, and Helias signaled to the professor with his chin, indicating that he should start. The counselor began to pace back and forth.

"What is your name?" asked the professor.

"Kathia." answered Athika, who relayed the answers as they appeared in Kathia's mind.

"Kathia, and then?"

"Kathia Four. But I'm registered as Kathia Cousins."

"Why did you betray your sister, Kathia Two, known as Bodieur?"

"I didn't betray anybody, I just do my job. Who can say which one's the traitor?"

Athika, seeing the professor's inquiring glance, added "She's sincere."

Helias, impatient, tapped the professor's arm.

"Where's Kathia Two now?"

"I don't know." And Athika nodded, confirming that the answer was sincere.

"When did you substitute her?"

"As soon as she left prison. Or rather, I left instead of her."

"What's become of her? Has she been harmed?" cut in Helias.

"One question at a time." chided Athika. "Anyway, she doesn't know."

"Who do you work for?" the professor resumed.

"I receive orders and I'm paid, but I don't know by whom."

"How do the contacts take place, by email?"

"No, physical mail and cell."

"What code do you call?"

"No code, it's a direct line."

"The postal address?"

"I only receive."

"What?"

"The credits."

"Where is your cell now?"

"Where it belongs. But I 'burnt' it when you captured me."

"So there's no way of tracing your 'employer'?"

"Not that I know of."

"How were you contacted the first time?"

"I received the cell in the mail."

"Why did you accept this assignment?"

"Because I had just lost my job. And because they told me it was a question of planetary security, like my previous assignments."

The professor was about to formulate another question, but the counselor stopped him.

"Excuse me, professor."

"Please."

"Where is Mattheus?"

"Which one?"

"'Our' Mattheus."

"In the capital, I think."

"Of Alkenia?"

"Yes."

"Do you have a base there?"

"Yes."

"Where, exactly?"

No answer.

"She knows, but she's been conditioned not to reveal it."

"Is there any way to get around the conditioning?"

"I don't think so. On the contrary, if we insist she's likely to lose all memory of it."

"Is Mattheus okay?"

"He's not been harmed, as far as I know."

"How many people, in addition to you three, are involved in this business?"

"Another four."

"Where are they?"

"In the capital."

"Where?"

"Don't insist. You'll just make her suffer, and it won't do any good."

"When was Mattheus taken away?"

"A bit before you arrested us."

"With the red diskette?"

"Yes."

"Did he give it to you?"

"We questioned him. For a long time. He was excellent. For a while he made us believe that he hadn't found it, in the place Helias told him about. Eh, Helias.... But at the end, with a massive dose, he blabbed. Right when that fox, over the cell, was sending me to the wrong place. They called me right away. By then, Mattheus had spilled everything, even where he had found the diskette. And so I discovered that he was taking me for a ride, and we went to him."

While Athika was telling them all this, they looked at each other in surprise.

"Why is she giving us all these details, without our even asking?"

It was Athika who answered.

"She's beginning to relax. Wait. Now she's thinking of Helias. Oh-ho! Wow! But what were you up to in the infirmary? Oh, no!"

And Athika looked Helias up and down, archly.

Helias blushed bright red.

"I.... I didn't know that.... In fact, that was exactly when I started to have doubts.... I mean to say.... Kathia usually.... But she, on the other hand..., she, how to put this, she got me..., the other way round. Well, you know, it was different. And anyway it's none of your business...."

Uproarious laughter. Laughter almost to the point of tears, in Athika's case.

"Wait, there's more...." And Athika went back to relaying the girl's thoughts.

"Having a good time? Me, not so much.... Why don't you wake me up and let's put an end to this? By now you've convinced me, anyway...."

"What's she saying? What's going on?" Helias, still red, asked Athika.

"She's, so to speak, coming over to our side. Don't forget she's also reading your thoughts. And, in particular, she read Helias quite a bit, starting much earlier…."

And here a wink at Helias, now redder than ever, was something Athika simply couldn't forego.

"Basically, it seems she's coming around to the conviction that she was on the wrong side. That 'the others' had misled her about you. Wait a sec…. Yes…. Yes…. Got it…. Sure, I'll tell them. As I suspected…. Major revelation…. It appears that the…, uhm, 'gifts' of our champion here had a decisive role in this."

This time Helias too, after burying his face in his hands, joined the general laughter. Only the secretary refrained from laughing as, all business, he continued to record, keeping his distance.

But the laughter died out almost immediately, as Kathia's first moans were heard.

"And now this, even….", tone-deaf, one of the 'nurses' lamely joked.

"She's not well." said Athika with concern.

"How's that? What's going on?" asked the professor.

"It's as if her higher brain functions are being blocked. She's losing consciousness, slowly."

The professor hurried to the gurney.

"Where's the cell? Where did she say she had it?"

"She was searched. It wasn't on her…." answered Athika.

The professor felt Kathia's neck.

"Damn! It's here. An adhesive strip, perfectly camouflaged."

And he tore it off.

"Check the other two for them as well. Remove them and take them away. Have them examined."

And two more strips were found and taken to the laboratory.

"Silence! I can still hear something…." said Athika.

"I knew it…, that there was a catch…. They've heard everything…. And now they're having their revenge…."

Then Athika added "She's fading."

"Kathia, listen! Give us a lead…." said the professor.

"Airships…. No, airport…. Roads…."

"Go on!"

"Nothing. I can't make out anything anymore. I think it was 'Airport Road'."

Athika stood listening for a while longer, then, complying with a request from the professor, she approached the other two gurneys. And shook her head.

"Nothing. Nothing here either."

"What does that mean?" asked Helias.

"Unconsciousness. Practically a vegetative state."

"But is it reversible?"

"Who can say? We'll get them to the infirmary right away. We'll let you know."

"But what happened?"

The professor answered.

"As far as I've understood, in the event of danger they were supposed to sabotage their cell. Or that's what they thought they had done. In actual fact, they had activated it in such a way that the people at the other end could hear everything and, if they saw fit, could also send certain signals, with the devastating effect we've just seen."

"Beasts!" exploded Helias.

"Let's go back." said the counselor. "To find out about this Airport Road. I've also got trusted men in the capital who can start investigating on site."

While others investigated, Helias and the professor returned to the office to take stock of the situation.

"So, in practice, while we know almost nothing about them, they know everything. Now we've let ourselves be spied on like fools. And before that, they must have interrogated Kathia in prison, to the point where they knew the smallest details and could trick us by replacing Kathia and Mattheus.... Professor, are you listening to me?"

"Yes, yes.... I was thinking.... Go on...."

"The weird thing is that it all began with a sort of favor you did for your pal Nudeliev. Then you got me involved in it.... And now we have a kind of war on our hands. With three people in a coma back there. And Kathia and Mattheus lost, locked up who knows where—and that's if we want to be optimistic, otherwise.... Not to mention my parents, since I haven't a clue where they might be. Pretty cheerful, the whole business. And we don't even know who we're fighting against, or what exactly they want. Now they've got the diskette. And so what do they do? They just about kill three people.... For what reason? Out of fear that we would get them back? We, who? A handful of odds and sods, working off our own bat and trying to prevent nebulous and unspecified future catastrophes which may or may not come to pass.... You know what they're going to call us? 'The daring defenders of the past of a future that's not yet present'. I'd say we're all crazy, stark barking mad."

"Have you finished?" asked the professor in irked tones, looking at him askance.

"You want me to tell you the truth, professor? If it weren't for all the people who are dear to me that are caught up in this mess, and probably in danger, I couldn't care less about your diskette, red or blue or whatever the hell it is. Who says it's got to be me to fix all this? What's it to me? What's my role in this business, and who got me into it?"

"Have you finished? Blown off enough steam?"

"Yes! Or rather, no! I'll see this through to the end. But only to try and save the people who are dearest to me. Don't ask me again to find or keep any damn diskettes. I'd rather swallow them whole. And anybody who wants anything else from me can go take a flying...."

"In fact. Nobody is asking you to."

"Ah! Good!... What? Why's that?..."

The professor grew pensive.

"Because they've won. They've got the diskette. And they'll keep it. We can't do anything about it."

"And so we're going to let them get away with it, just like that?"

"It's not a question of letting. That's just the way it is."

"I don't get it. What do you mean?"

"I mean we'll do everything we can to free our friends. But as regards the diskette and the time trips, there's nothing we can do.... Because they've already taken place."

Helias was unable to speak for a few moments.

"So we've lost. How did you find out?"

"It was one of the things I looked into, investigated, in these three years. I've told you only a very few of these things so far. What I told you earlier, about the mutants and the scientist, is little more than an outline of the events, things that I was only vaguely aware of before. I only tried to fill in whatever details I could, from the documentation I was able to find. But my real investigations were in another area. Which I'll tell you about after supper, if you like. If you're not too tired and you've calmed down by then. And if you're interested, obviously."

"I'm dead tired, and I need to eat, too. Maybe I'll feel a bit better afterwards. And I'll be happy to listen to you. Shall we have supper together?"

"Why not? Let's go. Hoping that there's still something left to eat in the dining hall at this hour. You see, that's the upside of having been found out, finally...."

"What is?"

"That we can trot happily off to the dining hall together, without worrying about prying eyes and eavesdroppers."

"Yes?" said the professor, answering his cell.

They had almost finished eating, in silence, each deep in his own thoughts, in the nearly empty dining hall.

"Ah, finally some good news...."

Helias was all ears.

"Yes.... I see.... But unharmed?... How did they find him?..."

'...find him...'. So it wasn't Kathia, then. Must be Mattheus.

Looking over at him, the professor mouthed, "Mattheus."

Helias's pleasure was tinged with disappointment. For a moment he had hoped....

"They're bringing him here?... Good.... Thanks for letting me know.... Yes, of course...."

Helias waited to hear the rest.

"They've found Mattheus, in a pub of some sort in the capital, near the road to the airport. Apparently he's fine, but in a confused state. Over-drugged, probably. They're bringing him here. Given his condition, they must have let up their guard, and he managed to escape and call...."

The cell again. The professor excused himself.

"Yes?... Please, tell me.... We'd better come.... What's she say?... Yes, we're coming immediately."

The professor explained, "It seems that Kathia Four is showing some signs of brain activity. It could be transient, so it's better to be there. A quick cup of coffee? I have the feeling it's going to be a long night."

"Yes, and with lots of sugar, please."

"What's she thinking of?" the professor asked Athika.

"Earlier, images were circulating, like single photos, with no apparent connection between them. Then some moving images, like interrupted film clips, but all very confused, like the beginning of a dream. Hard to describe, too, particularly since it all seems meaningless. They appear to be random images, welling up again in her mind. A cup with a teaspoon. A swing, swaying back and forth on its own. Other objects, places and landscapes, fairly ordinary ones. Few people, almost none. There's nobody even on the swing, though it's moving, but it's not the wind that's moving it."

"Can we try to ask her some questions?"

"I don't think she'll respond. I tried to call her by name earlier: no reaction."

"What's that screen? Some kind of electroencephalogram?" asked Helias.

"Something of the sort, but very advanced and sensitive. It separates the various layers of brain activity, distinguishing between the individual cortical areas. That's how we noticed the first images."

"I think there's a new signal, that wasn't there before."

Athika turned toward the screen.

"You're right. It looks like an attempt to open up channels of communication with the outside. Look here, the frequency of the neural dialog is increasing. See the change in the delta waves? And the propagation between perception and thought? Wait, let me 'hear' what's going on."

Athika stood 'listening' for a while.

"Try calling her…."

Helias moved next to Kathia and pronounced her name. Athika shook her head. Helias noticed that the professor was looking at his hands. Perhaps they were thinking the same thing. The professor raised his eyebrows and nodded. Helias reached over and stroked Kathia's hair.

"Kathia, can you hear me?"

A flicker of color on the screen.

"Go on! I don't know how, but she felt something."

Helias continued to stroke her hair, and to call her in his thought.

The flickering was there again, the signal coming and going.

"Something's there…." said Athika. "But I can't make it out. Because the images have also increased in intensity and frequency."

Then Helias said "Kathia, is there something you want to tell us?"

"She hears you! The images are quietening down. There. Now I can almost make out what's there."

Prolonged silence. Athika, eyes slitted with the effort, was concentrating on a far, faint signal.

Then she opened her eyes.

"Nothing more, now." she said.

"What did you hear? Did she 'say' something?" asked the professor.

"She repeated the same phrase several times. A kind of refrain."

"Tell us."

"Yes, but you should move away from her now, Helias. I think she needs rest more than she needs you, at the moment. I can see that you've helped her a great deal, but if she's going to get better, it will have to be a very long, gradual process, with nothing that could shake her up too much. You can understand that, can't you?"

"Certainly."

And Helias, after stroking Kathia's cheek one last time with the backs of his fingers, moved to the other side of the room.

"At first, I could only hear indistinct 'noises'. Then I began to make out a word or two, again and again. She was repeating the same phrase over and over, like a warning, or a maxim of some kind. She was saying 'What has already happened cannot be prevented. We can only work to make it happen.'"

Helias and the professor were in the 'projection room', the one where they had talked about the Feynman diagrams. The screen was on, and the professor was describing the findings of some of his investigations to Helias, illustrating them with images of the documentation.

The professor's cell rang. It was about Mattheus. They had brought him back. He was still in a state of confusion. They had run some tests on him and now he was being detoxed. He was sleeping, and wouldn't wake up for another half hour.

They continued their discussion for a while. Then they went to Mattheus.

He was still sleeping. Not completely detoxed yet.

Athika arrived almost immediately. She had heard them come in.

"He knows something." she said, smiling at Helias.

Helias smiled back. "About Kathia?" he asked.

Athika nodded. "She's alive. And doing well, physically at least. Or this, at any rate, is what Mattheus 'knows'."

"How did he find out?"

"That I don't know. We'll know tomorrow. Now it's better to let him rest. He's been through a lot. Worse than we thought. What they did to get him to tell them what he knew reduced him to a wreck."

"And the others?" asked Helias of the three remaining patients.

"Fairly stable. Kathia is the only one who continues to show signs of recovering."

Helias and the professor wished each other good night. Both were off to bed. Both needed to sleep.

Once in his room, Helias turned on a low, diffuse light and went into the bathroom. He urinated copiously. Then he looked at his wound in the mirror. It was practically healed, just a slight scar left. A nice hot shower, that's what he needed. He undressed and entered the stall. But stopped immediately, before turning on the water. He had heard a noise. Like a chair being moved. He stood listening. Nothing.

He got out of the shower and went to check that he had locked the door. But the dim light and his fatigue prevented him from seeing into a dark corner, where a shadowy figure was waiting, motionless and in silence.

Ten minutes later he exited the shower, clean and dry, and walked toward the bed. But out of the corner of his eye he saw something unusual on the desk. He came closer. It was a sheet of paper, something written on it in a tiny hand. He could have sworn that before, when he came in, it wasn't there.

He turned up the light and took the piece of paper. A shiver shot up his spine, as far as the nape of his neck. Hand trembling, he brought the sheet closer and read the first lines carefully.

"Since you've certainly recognized the handwriting already, maybe it's better for you to take a seat and calm down before going on."

Like an automaton, Helias obeyed and sat, never lifting his eyes from the paper. He reread, incredulously, the first sentence.

He looked up then, scanning the room around him. No, nobody there. And the door was closed, he'd already checked. But, this, clearly, was not a significant detail. And the noise he'd heard before made sense now. He drew a deep breath and sat thinking for a while, staring blankly into space. Then he went back to reading.

"You'll think it's stupid, but you have no idea how happy I was to see you again. How is the injury doing? It should be pretty much healed by now. Yes, and some of the credit for that is mine. When? When you were asleep, under the boulder. Before they came to get you. Yes, I have them myself, the diskettes. I'll give them back to you when the moment comes, when you're no longer in danger of being hunted for them. Now relax for a while. Then come outside, to the lakeshore. There's nobody around at this time. We can have a chat, if you like. See you soon."

Helias's hands were cold and clammy. Once again, he read those words, with meticulous attention. To be certain he had understood, understood absolutely everything. Then he raised his head and closed his eyes. He could feel his heart hammering, his temples too. There was a sour taste in his mouth, and his throat was dry. He was needed to drink something, but he didn't move. The words echoed in his mind, and it all seemed incredible still.

Finally he rose, drank, dressed, slipped on his jacket and went out.

The night was cold, but brightly lit by Nasymil. Other stars shone, uncounted, and the only sound was the soft steady plashing of the waves.

He was walking along the lake, deep in thought. When he heard footfalls pattering behind him. Afraid to turn, he slowed his pace, finally coming to a halt while facing the lake. The hooded figure came up beside him.

"Everything all right?" it asked him.

He nodded. That voice gave him the shivers. Helias made as if to turn.

"No, don't look at me. It's better not to."

And the figure took half a step backward, standing almost directly behind him.

Helias was panting, as if he had just stopped running.

"What do you want? Why have you come here? Why can't I look at you?"

"You already seem agitated enough as it is. Let's calm down and talk, okay?"

"Okay, okay. You're right, sorry."

"I know there are a lot of things you'd like to know. Unfortunately, though, I can't tell you everything. But maybe there's something I can reassure you about, anyway."

"What's become of Kathia? And of… of my… parents?"

"They're all right, now."

"They're prisoners?"

"Yes, they are."

"What can I do for them?"

"You'll know when the time comes. You'll see."

Then Helias smiled. The tension was beginning to evaporate.

"And you, how are you doing?"

"Not too bad. As you can see, I'm still alive."

"I can see that. Or rather, I hear it. It's a weird feeling, hearing your voice."

"The effect on me is more or less the same."

Helias laughed.

"You know you're just the same old piece of shit you always were?"

The hooded figure laughed too.

"Look who's talking!"

And at that, they both joined in a resounding peal of laughter.

"If it wasn't for that hood I'd give you a good slap or two."

"That's exactly why I'm wearing it."

"No. I'll bet you're wearing it because you've got so ugly now that you're ashamed to let yourself be seen."

"Hey, young fella. A little respect for your elders."

Which made them break into helpless laughter again.

"Somebody's coming." said Helias.

"I know. We'd better split up here."

"Already? Just when I was beginning to enjoy myself?"

"Yes, it's better."

"Will I see you again?"

"Probably."

"Tell me something more."

"You've already heard what you need to know."

"And what would that be?"

"What has already happened cannot be prevented. We can only work to make it happen."

Helias brooded over this for a while.

Faint footsteps roused him from his thoughts.

He would have liked to have asked how much time had gone by.

But when he turned, the dark hooded figure was already far away.

Chapter 14
The Meeting Had Left Helias Kadler Shaken and Confused

The meeting had left Helias Kadler shaken and confused. He lay on his bed, unable to sleep despite his fatigue. Thinking that, whatever dream he might have that night, it could hardly be more unreal than everything that had happened to him lately.

And it was exactly that sense of unreality that had relegated him, unwillingly, to a role that had almost always been passive as these events unfolded, surprising and unpredictable events. It was like being catapulted into an unknown world, governed by its own strange and incomprehensible laws. Exactly what had happened to him, since setting foot on Alkenia.

And now, with that meeting, he had reached peak stupefaction. It had finally put him on the ropes, punch-drunk and battered, but in a way, he felt, that gave him the strength to battle back, just as a prize fighter leans on the ropes, stretching them as far as they will go and then bounces again into the ring, fists flying. And he couldn't help but think of that dream, the one with the Martians on the elastic cords, in one of his first nights on Alkenia. The dream where he, too, ran and jumped on the cords, higher and higher, like the others. Even if he didn't know what the purpose of it all was, he had learned to jump too. If for no other reason than not to slip and fall, tumbling down into the darkness, who knows where. It was necessary to jump, in that utter and absolute blackness. Even if he couldn't see anything, even in his fear. Even if he didn't know whether those jumps, each higher than the time before, each more harrowing than the last, would ever come to an end.

And the same feeling of adrenaline coursing through his veins that evening prevented him from sleeping.

Then he thought again of Kathia and his parents, whom he knew were alive and well, somewhere. But where? Held prisoner by whom? And though the first thought was calming, the questions it raised unsettled him again, and led to more questions, new ones posed for the first time. And at the same time, he felt reinvigorated by the thought of Kathia and his parents, as if they were waiting for him, because they needed him. And who, if not he, had a chance at saving them now?

And in Helias's mind, a mechanism hove into motion, searching for questions, spontaneously, continuously, almost without his realizing, questions without

M. Villata, *The Dark Arrow of Time*, Science and Fiction,
https://doi.org/10.1007/978-3-319-67486-5_14

answers but which together, at the end, would form a picture. Not a picture of questions: the picture itself would be the answer. Like putting together the pieces of a puzzle, pieces that earlier, individually, seemed meaningless. But you had to have the right pieces, ask the right questions. The answer, the picture, would come on its own. So, long ago, it seemed, the professor had explained.

It was far into the night before Helias finally slept.

But the picture had not taken shape. Not entirely. And when, the next morning, Helias woke, he had the feeling that some of the questions and conjectures that had drifted through his drowsy mind during the night were not nearly so realistic now, in the light of day, as so often happens to those thoughts and meditations that arise in darkness and solitude, when the mind seems free from distractions, but which melt away in the morning, as they come up against reality. But he wouldn't have been able to say which was right: the undistracted thought, undistracted but perhaps distorted by sleep, or what we understand to be real, but which in turn could be no more than a distorted vision, whose features seem real only because we are accustomed to them, and have come to like them. He didn't know, in other words, whether true clarity had come with the morning, or before sleep. And so he had that unpleasant sensation of not being able to tell what was real, and what wasn't.

To make his mood worse, there was the disappointment of realizing that he couldn't remember certain elements, certain questions, that had seemed so essential the night before. The only thing that remained was the feeling of having made a discovery, or an intuition, of fundamental importance. And so even what little was left of his unfinished picture had fallen apart, completely and finally fragmented.

But he decided not to take it too hard and, as he went through the bathroom door, reflected that the pieces, the questions, were all there, somewhere, still working away in his mind.

Helias was finally able to relax when he went to breakfast. Immediately beforehand he had phoned the professor and made an appointment with him, for half an hour later.

Now, as he ate, he could feel that the mechanism was slowly starting up again, and he was careful not to jam it. He was having that 'stubborn grease' feeling of his, like in those laundry detergent commercials where the stains loose their hold on the grimy fabric, clumping together into fat black globs that float off and away, leaving the wash whiter than white. They were the questions of the night before, coming together again, retaking their shape and extricating themselves from the sleep-clogged meshes of the mind. Then it was like when droplets of mercury are evenly distributed over a soft surface, but all it takes is a moment's pressure, the slightest pressure, and two droplets come together and coalesce. And then a third and a fourth. Gradually the droplet becomes a drop and then a ball, the soft surface curves under its weight, and all the other droplets roll down to joint it. Like the birth of a star, when all the surrounding gas is attracted and builds up around that globule that, at the beginning, was only slightly denser. Until the pressure of the gas that continues to accumulate becomes so great that it triggers the first thermonuclear reactions. And the ball of gas becomes a star. And the star shines with its own light, starting where once there were only solitary shreds of cold, rarefied gas.

And a picture took shape in Helias's mind, starting from isolated, insignificant globs of grease and droplets of mercury.

The professor looked curiously at Helias, noting that strange glint in his eyes.

"You're looking refreshed. Good. Tell me what's on your mind…."

"There's one question, first, that had already occurred to me but I hadn't attached too much importance to it. Now, however, it strikes me as crucial. I'm talking about three years ago, when Professor Nudeliev's computer system was hacked into. Why would our hacker, whoever he was, have tried to steal the software when the product wasn't finished yet? Why not wait for the final, definitive result? What good to him was a program that was incomplete and unusable, at least as far as was known, or should have been known outside of the project? We know now that it was already operative, though still at the experimental stage, but who could have known that at the time?"

"You're thinking it was an inside job? That crossed my mind too. And not only mine. And there were extensive investigations that also—and especially—looked into that. But without turning up anything: everybody involved in the project was cleared. Or, at least, nobody could be suspected more than any of the others. And nobody had a serious motive, or the opportunity to hack the entire system…."

"And yet, hacked it was. And it's certainly less difficult from inside. As for the motive, maybe just plain ordinary corruption: somebody on the team could have been paid off by somebody on the outside who wanted the software and had the wherewithal to use it…."

"Yes, I was getting there. No, there was nothing of that kind either. No questionable dealings with outsiders, especially with the few suspects who might be interested in a product like this…."

"Hmm, as I thought…. And so…."

But Helias broke off in mid-sentence, ruminating over something.

"And so?" echoed the professor, looking at him dubiously.

"No, nothing, just an idea of mine. But it would be premature to talk about it now. What I'd like is to be able to have a look at whatever records were kept of the recruitment procedures for the team that developed the software. Assuming that there was an open call, as I imagine there was…."

"Yes, there was. Though as is often the case in this kind of thing, it was—at least to some extent—little more than a formality. In fact, many of the team members had already been working with the professor for some time, before the project was funded and the open call was announced. Some, however, were new and were recruited for the first time through the call. Clearly, some of the investigation's most intense scrutiny focused on them…."

"How many were there on the team, in all?"

"A dozen, I think. Of whom over half, I'd say, already worked with the professor. In any case, I believe the records have already been looked at quite carefully in the appropriate quarters…."

"I'd like to see them anyway, if possible. They may not have looked for what I'm trying to find."

And the professor went back to looking at Helias quizzically, wondering what was brewing in that head of his.

"I can try to ask the counselor. He should already be back on Thaýma now. He's got the clout to root out just about anything...."

"Yeah...." replied Helias, again lost in thought.

"I'll send him a message straightaway." said the professor, turning to the computer and beginning to dictate.

"Wait a minute.... Don't send that yet. Ask him for information about a certain Rosa Stawinski, born on Thaýma around ninety years ago."

"How's it spelled? I can try to do a search myself, for that...."

Helias spelled the name and the professor typed it in the search box. No hits. Just a namesake in her forties.

The professor sent the message, with the two requests. And Helias marveled at how that message would be sent backwards in time and travel for more than a year and a half before reaching its destination on Thaýma. Where it would be answered, and the answer would take the same amount of time, but in the opposite direction, and would reach them slightly after their first message was sent.

And so it was, in fact. The secretary had answered, after calling the counselor who was away from the office. He had attached the records, and now Helias, with the professor translating, pored intently over the pages.

At a certain point he let out a sigh and sank back into his seat.

"Found something?" asked the professor.

"Yes and no. New questions, more than anything. Look at these four names here: we need to find out what their jobs were after the call.... As I suspected, perhaps instead of looking at the applicants who were hired, it would have been better to focus on the ones who weren't. Look here: all four had worked with Nudeliev before, and they failed to meet the cut by the barest margin, but if you look at their vita and so forth, they certainly don't seem any less qualified than the successful candidates...."

"Maybe they screwed up on the exam or fluffed their interview...."

"Yes, in fact it was the interview that was their downfall. If they hadn't made such a bad showing there, they would have been among the top picks. And yet you told me that it should have been little more than a formality. And anyway, what weight can an interview have, since Nudeliev already knew them well? Everything makes me think that Nudeliev didn't want them, despite their long-standing work with him and their professional ability...."

"But Nudeliev wasn't the only one assessing candidates at the interview, there were at least some other members of the committee...."

"Well, all your friend had to do was cook up some questions he knew they wouldn't be able to answer, deliberately sinking them...."

"But why?..."

"That's something we can ask him directly.... But first we've got to know where they were employed afterwards."

"So you're thinking there was a sort of vendetta, engineered by people who knew exactly what to do, who knew the systems and the precautions that the professor usually adopted perfectly well, and perhaps jumped at the chance to sell themselves to potential competitors?"

"Who knows?..." laconically answered an Helias who was now more pensive than ever.

With the professor who was beginning to be irritated by this uncommunicativeness. And by this reversal of roles, where now it was the 'young man' who was calling the shots, and he who was supposed to be hanging on his every word.

Helias had a quick look at the second attachment, asking the professor to translate a couple of words he didn't understand. There weren't more than a dozen lines and, while skimming through them, Helias nodded to himself from time to time.

Then came a moment when the professor felt he could regain his rightful role and turn the tables again. It was when Helias asked "Do you know why my parents, two earthlings, worked for you Thaymites?"

The professor, with obvious satisfaction, assumed his lecture-room manner and drew breath for a long, deeply self-indulgent answer. But then he realized that he didn't have any particular answer to give, or at least not the one Helias was after, and, disappointed, tossed out an anything-but-academic "Well, I wouldn't know..., sometimes earthlings with specific abilities are invited to collaborate with....". But he broke off, because Helias, caught up in his own musings, wasn't even listening.

Exasperated, his words came out almost as a shout: "And you, do you know? Say something, for god's sake. What's running through your mind? Out with it!".

Helias slowly raised his eyes and turned toward him, teasingly.

"You will know in due time." he said, sententiously but with a smile.

So there! It was his own little way of getting his own back, against all things Thaymite.

At that point the professor, grumbling under his breath, had left the office for the adjoining room. Where the refrigerator was. The fridge that, judging from the noises Helias heard, had been expertly raided by the time the professor returned a few minutes later, visibly more relaxed but still determinedly sulking.

Then the professor, without saying a word, had set himself to composing the new message with the request for information about the four unsuccessful candidates.

The answer had come after several silent, interminable minutes. The professor had read it and, still without speaking, had turned the screen toward Helias. Once again, it was the secretary who had answered. Nothing. No record of where the foursome had been employed subsequently. Nothing, for all four of them. Hardly likely to be a coincidence.

Then Helias, remorse vying with annoyance at the professor's pouting silence, had decided to tell him his theories, while leaving out a few decisive details that he wasn't sure about yet. Essentially, he had given him a general outline, something they could discuss and that was still open to various interpretations, but without revealing his own conclusions.

Initially, the professor had wrinkled his nose and made objection after objection. Then, as Helias continued to explain his reasoning, he had—however grudgingly—been obliged to concede that Helias's argument at least held water, and that there must have been a modicum of truth in it.

Together, they decided to make a visit to Professor Nudeliev, to ask him a few questions and see how he reacted. Or at least the professor was left with the impression that the decision had been made together. In reality, the visit had been in Helias's plans for quite some time.

A back-now for Thaýma was scheduled for half an hour later. They would thus be boarding in the afternoon. They had had themselves put through to the counselor on his direct line, to inform him of their arrival and of the planned visit to Nudeliev, but without providing further details.

There was plenty of time for lunch, which they ate together in the dining hall, and for Helias to have a long walk on the lakeshore, where he gave his memories free rein. Memories of Kathia, of his parents, of what had happened to him since he first arrived at the Kusmiri Center, down to the unsettling encounter of the evening before. Then, before the final preparations for the trip, Helias spent half an hour at the computer, searching the web. Afterwards, he opened one of the desk drawers and sat contemplating the long, curving object half hidden among the sheets of paper. It took him a while to make up his mind. Then he checked the charge.

Chapter 15
In that Moment, Helias Could Remember Very Little of His Conjectures

In that moment, Helias could remember very little of his conjectures about Professor Nudeliev, of his discussion with Professor Borodine a few hours earlier, and of the questions he had planned to ask the project leader.

He had the same feeling he usually had before the exams. When, after prolonged swotting—and a certain amount of time trying to second-guess the questions and picture what the exam would be like—the hour would come, and with it, the realization that nothing was the way he had imagined it, and the only thing left to do was empty out his mind, a clean sweep, and listen to the questions, the real ones, concentrating on the present, the here and now, the only present that counts.

But he didn't understand why he should feel he was sitting an exam; after all, he hadn't come to be quizzed, but to do the quizzing. Then he saw that it was in fact his own answers that had to be put to the test, the answers he had already formulated in his mind, and that the real question now was how these answers measured up against the yardstick of reality.

Or maybe it was also the fact of being face to face with an eminent professor and scientist, and a Thaymite to boot, who was now regarding him with a steady proctorial eye, exactly as if he were the hapless examinee. And he was so different from how Helias had imagined him. So different from the professor, or rather from George, as he called him.

"How are you doing, George? You're looking well. If you had let me know earlier that you were coming, I could have organized something…"

It sounded a bit like a reproach, and Helias wondered if the two professors had always been so formal with each other, or whether the formality was because of his presence.

"You're looking well too, Valeri. I apologize for practically bursting in on you like this, but it was decided on the spur of the moment, and it's not just a courtesy call. Dr. Kadler and I urgently needed to talk to you about certain matters concerning…"

Whereupon Valeri Nudeliev, murmuring some polite formula that Helias didn't understand, once again looked the young visitor over and shook his hand.

M. Villata, *The Dark Arrow of Time*, Science and Fiction,
https://doi.org/10.1007/978-3-319-67486-5_15

The two professors were speaking to each other in European—Nudeliev, who was clearly less accustomed to the language, a bit stiltedly—so that Helias would not feel left out.

They were in the scientist's lavish residence, overlooking the sea. A sprawling one-storied villa, with the full complement of swimming pool, gardens and miscellaneous Thaymite amenities which, to tell the truth, were very much at odds with the austere figure of the professor. That was the biggest difference between Nudeliev and his colleague Borodine, as well as from the person Helias had pictured, aside from being a few years younger and very much thinner: that austere aspect, a bit like that of the counselor, with his deep-set, inquisitorial eyes.

Perhaps all that luxury and all those comforts were more a reflection of the rest of the family. Because Nudeliev had a family, even if the place seemed deserted at the moment. No sign even of servants, though a setup of that size and scale clearly didn't run itself.

After a few more pleasantries, Professor Nudeliev invited them to follow him to his studio, where he evidently felt more at ease.

Here Borodine gave him a short summary of the latest developments, which Nudeliev, enthroned behind his massive desk, received without emotion, clearly having hear some of this before.

Helias was seething with impatience. Not so much for the questions he wanted to ask, which kept surfacing in his mind, but for that casual, even rambling, tack that the conversation between the two professors had taken, all while he had an exam to get through, and mounting nerves. And not least because of the glances that Nudeliev continued to darted at him.

At a certain point, he took advantage of a brief lull to join the conversation. Or rather, he broke into it, with the question that was uppermost in his mind.

"Professor Nudeliev, why according to you was your system hacked three years ago, when the product of your research was, as far as was known at the time, not ready yet and…"

As was to be expected, the atmosphere in the room turned decidedly frosty. Helias saw that George was looking at him with distaste, for having so cavalierly kicked over his whole, carefully constructed diplomatic edifice. Nudeliev shot a look at him, somewhere between surprised and pleased, but otherwise unperturbed.

Borodine was about to speak, but Nudeliev stopped him with a raised hand.

"Right. Let's lay our cards on the table, then. Sorry George, but I was getting a bit tired of all this circling around, too. Much better to speak openly. I'm no fool, and I had figured you were here to ask me precise questions, because in one way or another you have doubts about my conduct. As for your particular question, Dr. Kadler, my most obvious answer would be that I know no more than you do, or, if you like, that I have no idea."

Helias had rarely heard a more sibylline answer. It was clear that Nudeliev was waiting for more questions, to find out how much they knew, or had guessed. And Helias did not make him wait long.

"Professor, when did you involve Simona Villardo, daughter of Rosa Stawinski, a native of Thaýma, and of her earthling partner, Marko Villardo, in your project?

And why did you involve her and her husband, both earthlings, in a Thaymite project?"

"I didn't know that you were aware of the origins of your family, Dr. Kadler…."

"I wasn't, in fact. As is obvious, I was kept in the dark about them. But, as you can see, I found out on my own. I just connected the dots between the fact that my parents worked with the Thaymites, and the haziness about my grandmother that prevailed in my family. But answer my questions, please."

"Certainly. Your parents knew of your grandmother's origins and thus knew of Thaýma's existence, and, from a scientific as well as a personal standpoint, were obviously fascinated by it. So it was by no means difficult to recruit them for a Thaymite scientific project, especially such an intriguing one. They both had excellent minds and, even more to our advantage, were already working at the Martian station, which was ideal for our experiments for a number of reasons, including its distance from Thaýma and the fact that being run entirely by earthlings put it beyond suspicion."

"So you were already at the experimental stage, top secret."

"In fact. We started the experiments that involved them a little less than a year before the events that led to the project's discontinuation."

"What kind of work are you doing now, Professor?"

Nudeliev studied Helias at length before answering. And, with the conversation turning into a grilling, Borodine, too, gave him another look of reproof.

"It seems to me that we're straying off the subject…. Anyway, nothing special, since then I've gone back to my academic commitments and am mostly concerned with theory, again in connection with time transfers."

"When did you hire Kathia Cousins for the first time?"

This time Professor Nudeliev shifted in his seat, at least, blinking repeatedly.

"A little before the Kadlers were brought on board. We could trust Simona, if for no other reason than her Thaymite descent, but we had no guarantees that your father would accept the job and keep the secret. So I sent Cousins to Earth, at a time when your father was there, to 'spy' on him, in the guise of a graduate student who was looking for a faculty advisor…."

"So that's why my father recognized her after he was wounded on Mars, or rather, he mistook the other Kathia—Kathia Bodieur—for her. But that doesn't explain why he feared her, or at least regarded her with suspicion. Unless there's more to it. Unless he knew that that person was working for someone hostile to him…."

A long pause followed. If Nudeliev's aim was to find out how much they knew, Helias's tactic was to feed him that information piecemeal, making him tense enough that sooner or later he'd make a mistake. And to some extent the tactic was working: Nudeliev had forgotten to protest when Helias had said 'for the first time', implicitly admitting that there had been a second time, when, recently, Kathia Cousins had been hired to impersonate 'his' Kathia.

Helias felt that the time had come to up the ante and, almost automatically, his hand slid slowly toward the side pocket of his trousers, hidden from Nudeliev's sight by the desk between them.

In the pregnant hiatus that followed Helias's last words, his voice was heard again, in measured, neutral tones: "Do you know a certain Geremy Stuerz, Professor?"

Perhaps Helias was expecting a simple 'No.' But the professor turned a long penetrating gaze on him. Then, leaning back in his chair and without taking his eyes off him, said "Go on, please."

"I've known him for a long time. Not well, but for a number of years, at college. I wouldn't be able to say exactly when I met him. But Professor Borodine made me think yesterday, when he said that perhaps Stuerz, too, had been sent to Earth to watch me, after the message was found that my mother left at the station at the time of the attack, before disappearing together with my father. But something didn't add up. I remember at least one occasion, before my parents' disappearance, when he was hanging around. So he must have been there for another purpose. That was the three-month period between the hacking and the attack on my folks. Let's add another detail. When we were at the Martian station, the pilot, who along with Spitzer was later found to be an imposter, at one point told the captors not to harm my parents, and that they would be punished if they did, because my parents were of more use safe and sound. It almost seems as if the attack was more of an attempt at kidnapping than cybertheft. In which case: why did my parents have to be kidnapped? So that they could continue to work on the project, as the only ones who had certain essential information? That could be, though from what you told me before, it doesn't seem likely…."

Helias paused, watching for a reaction from Nudeliev, who to all appearances was listening calmly, almost reclining in his chair.

"Perhaps they thought that the Kadlers would have been able to reconstruct the compiler, since they must have known it very well…."

"I don't think so. Actually, this whole story of recovering the compilers reeks of red herring. At least at the beginning, up to the point when it was decided to get them back. Not, however, to use them, but to prevent them from being used by others later. So Kathia Cousins said when they had drugged me to the gills. I think my folks were kidnapped mostly to keep them from talking, and revealing what they had found out. And this is where my father's second meeting with Cousins comes in, when she read his intentions in his mind, and he understood that she was working for the enemy…."

Another pause. Because now the professor had sat upright and was reaching for something behind the desk with one hand, probably trying to slide open a drawer. Helias slipped his hand into his pocket, wondering whether the weapon would work through the wood of the desk. But he had not finished formulating the thought when Nudeliev said "Easy there, Dr. Kadler. Leave that weapon alone, there's no need for it. I'm only getting a cigar, see?" And he raised his hand to show the cigar gripped between forefinger and thumb.

"I rarely smoke. And only outdoors. My family can't stand the stink. But this is one of those moments when only a cigar will do, even if they're sure to make a fuss when they get back. I suppose I'll have to air the place out before. If you don't mind…."

Nudeliev lit up and, after several voluptuous draws, visibly unwound. Borodine, though still at a loss for words, stopped squirming, as he had been doing for the last several minutes.

"Very clever, Dr. Kadler. My compliments, really. I think, though, that you've made a few errors, perhaps marginal for you, but not for me. If you prefer, you can go ahead with your reconstruction. Or you can let me talk, now. But first, tell me a couple of things. I imagine you're recording everything, aren't you? And soon someone will be arriving, or you will call someone…."

It was Borodine who answered, a bit regretfully.

"Yes, we're recording. And we informed the counselor of everything, shortly before coming here. If you prefer, we can call him right away…."

"There's no hurry. First I'd say that it's better that you hear my side of the story. Yes, Dr. Kadler, I bear some of the blame for what happened, but, just like your father, I can consider myself a victim too. You've presented a valid reconstruction of the events, but you've got your target wrong…. If I were to ask you, Dr. Kadler, what is the most precious thing in our lives, what would you say? Life itself, perhaps? Or the people we love? But there's something that neither you nor my friend George can image, something both your father and I have. Children, Dr. Kadler. If I had known how important children would have been in this whole business, I would never have asked that my project be approved. Yes, your father was blackmailed, and Geremy Stuerz was on your tail, ready to strike if your father refused to cooperate. And he wanted to refuse when he realized how far beyond the pale those experiments were, he saw the dangers inherent in them. But in my blind lust for knowledge I insisted that he continue, and I invited him to a meeting where Cousins was also present, but made up so that he couldn't recognize her. But I wasn't the one who blackmailed him, I'm not a criminal. Shortly thereafter, I received an anonymous letter threatening my children's safety unless I agreed to the letter's demands. And what was demanded of me was nothing less than the project. I was told to let it be stolen, or rather, I admit, to be an accomplice to the theft. And I did, even though I could well imagine what consequences that product would have in the wrong hands. And the rest was all an act on my part, a sham, starting from the compiler in your possession. A red herring, as you put it, they already had everything. But they didn't have the knowhow, and so they kidnapped your parents. I'm not looking for excuses, I sold out. But the stakes were too high for me: my children's lives. And I had to go on collaborating, a nightmare…."

Nudeliev paused to stub out his cigar, eyes glistening.

"You might not believe it, but you've done me a favor, and I'm relieved. I'm glad you're here. Maybe the nightmare is about to end…."

He seemed entirely overcome, slumping in his chair and apparently sapped of all willpower.

"And I was the one who put them on your trail, when they decided to recover the diskettes. For that, I apologize to you, however little good it will do…."

Then he roused himself, as if he had decided to take the bull by the horns.

"But I can try to make amends for the damage I did to you, at least in part. But we have to be quick, before they wake up to what's going on. They're surveilling me…. Dr. Kadler, I know where your parents are. We can free them, if we hurry…."

"We'll call the counselor…." said Borodine.

"No! They could intercept the call. And make all the evidence disappear, Kadlers included."

"He's right…." said Helias. "Let's go!"

They went down to the basement, where Professor Nudeliev opened a well-concealed secret passage. After a long, narrow corridor, they turned into a dimly lit cross tunnel, where a four-seater cart was waiting. Helias sat facing the two professors and the cart took off, bowling down the tunnel.

In the weak reddish light, Helias observed the two faces in front of him, so different from each other. No one spoke. Apparently only Helias was feeling the tension of that moment. Nudeliev had his eyes closed, and seemed almost to be dozing, as if he was resting up for whatever would happen next. Borodine, too, looked quite relaxed, probably relieved to have learned that his old friend was more victim than criminal, after all. And maybe he was thinking that everything was going to turn out well.

Helias, though, was struggling to keep his calm and breathe normally. But his blood was pounding in his temples and his throat was parched. Especially, he could feel that hard object in his pocket pressing against his thigh, as if nearly all his senses were concentrated on that point. He could even feel the veins pulsing beneath it.

After a few hundred meters the cart slowed and came to a halt opposite another corridor at right angles to the tunnel. They got out and walked down it, Nudeliev leading the way, reaching a door that opened onto yet another corridor, this one wide and well lit.

As they hurried down the passage, Helias lagged a bit behind so that they wouldn't see the sweat beading his brow. Then he slowed down and put his hand in his pocket.

"Professor, I'd like to ask you another question…."

"Not now, we've got to be quick, we're almost there…."

"No! Now. Professor."

Nudeliev stopped and spun around towards Helias. Who was pointing his gun.

Borodine stopped too, looking at them uncomprehendingly.

"What happened to Gasler, Thomay and your other two former co-workers who you deliberately kept out of the project?"

The scientist regarded him with a strange smile.

"I was expecting this question, sooner or later…. They're down there, where the tunnel leads, in the lab where they've always worked for me, on my real project. The project you know about, the official one, was little more than a cover. Or rather, it's where I produced the preliminary results, which I then perfected and used here. To keep my real results and aims under wraps. I picked the four of them because they were the most unscrupulous, and they would have been faithful to me without qualms as long as the pay was good. Yes, Dr. Kadler, you guessed everything. I'm the mind behind all this, no blackmail and no anonymous letters. My compliments, again. I'm just a bit surprised that you let yourself be taken in by my little sob story, even if it wasn't for very long…. Ah, perhaps I hit all the right notes: the tears in the eyes, deliberate tears caused by cigar smoke, the raised hopes of freeing your parents…. Yes, I must say it was a masterful performance. Even though I knew you had every reason not to believe me…."

"In fact, I didn't believe you, not for a moment. I only wanted to see what you were getting at, and if you really would have taken us to my parents. But they're not here, are they?"

"They're not far away. But out of your reach. I had to kidnap them, they had understood everything. Like you…."

"Professor, call the counselor. Let's put an end to this."

"No, I'm wrong, you didn't understand everything. You didn't understand that my purpose was to bring you here: you're trapped."

Borodine was fumbling with his cell.

"What do you mean, trapped? Professor, what are you waiting for, why don't you call?"

"There's no coverage…."

"In fact, we're isolated here. As I told you: you're trapped."

"Are you perhaps forgetting that I'm pointing a gun at you and ask for nothing better than a good excuse to shoot you?"

"And then? Who would get you out of here? The door down there is secured and you don't know the password…."

"Well, if you'd rather try writhing in insupportable pain, I can oblige you right away, but I don't recommend it. Let's go. Get moving!"

But Professor Nudeliev didn't move, twisting his lips in a malignant grin as he looked over Helias's shoulders.

"Freeze! Dr. Kadler. Drop that gun."

A voice had rung out in the corridor behind him. A voice he had heard before.

Chapter 16
But Dr. Kadler, that Hot Afternoon Near a Sea on the Planet Thaýma

But Dr. Kadler, that hot afternoon near a sea on the planet Thaýma, was angry. Really angry. Nobody, but nobody, was going to tell him to freeze, or order him around like that, tell him to do this, drop that. No. But he also knew that he had no other choice, with a gun pointed at his back, however fast he might be.

And so he decided on a compromise. He lowered the gun, but didn't drop it. And he turned slowly, to look the unknown aggressor in the face. Enough time to recognize him, and raise his gun. Then he blacked out, caught in the crossfire between the newcomer and Professor Nudeliev.

"Welcome back!" exclaimed the latter, sneering at Helias who, hands tied behind the seat back, was coming to his senses in that lab or whatever it was.

"I believe you know each other already, but allow me to introduce my dear business partner, as it were, Mr. Petersen, who is also the personal secretary of our beloved counselor.... As you can readily see now, I was well informed of your visit as well as of your 'investigations', and my friend here has always been an invaluable source of information about everything that went on between you and the counselor. Including the unfortunate interrogation of Kathia Four and the rest, when he was ready to intervene with a special device planted in her camera, in case anything went wrong with the 'deactivating' cells...."

"You make me sick. Both of you. Fine pair of shits...."

"Now, now..., insulting us will do you no good. Actually, I think we make quite a good pair. You see, he knows nothing whatsoever of science, no offense, naturally. You know what he's interested in? He's an ambitious sort, not cut out to be a mere secretary. I saw that the minute I met him, and I suggested letting him in on the deal. All he had to do was tell me everything that came to his attention, and lend a hand when needed, in choosing a pilot and things like that for instance.... In exchange for what? Money, of course, wealth and power, to be stored up on Earth, where eventually he'll move. As you can imagine, if you're not squeamish and have the right tools, it's not hard to make money with a few time tricks.... Come on, George! Don't look so priggish, you've no idea how much all this costs me...."

M. Villata, *The Dark Arrow of Time*, Science and Fiction,
https://doi.org/10.1007/978-3-319-67486-5_16

"You're a madman…. You're running roughshod over everything, human beings included…. And for what, Valeri?" said Borodine, also tied to a chair.

"For what? For the greatest discovery of all time. Which is well worth a little sacrifice of insignificant human beings now and then…."

"Bastards!…" growled Helias. Which earned him a backhanded slap from Petersen, bloodying his nose.

Valeri Nudeliev was now so busily extolling his research that he seemed oblivious of the incident. And of everyone and everything, except himself.

"…You know what's always fascinated me most? The seventh dimension, and the opportunity it gives for changing history. Yes, George, I know you don't believe that, but as good scientists we have to let scientific experimentation provide the answer…. And this is where you come in…. You've arrived at the right moment, you know? Two human guinea pigs are just what I needed. With rats I didn't have much luck, in the last experiment…. You already know that I'm working on space-time transfers with no receiving station, or in other words, on the possibility of traveling freely in time…. So that one could go anywhere, back in time whenever you like, if 'inverted' at departure. But then you would have to invert again, if you want to participate in the past events, and not just be there, watching them run backwards in time. To be able to change history, in fact. Or even just in order to survive, or to return to your own time. And that also calls for a free process of time inversion that can be carried out anywhere, outside of the usual labs. This is what I'm working on now, and I've got to the home stretch, to the crucial experiment. Before, with the guinea pigs, I made a mistake and they vaporized on me, vanished into thin air, or rather, transformed into pure energy. But I corrected the error, and now it's your turn for the definitive test…."

"And what will you say when they start looking for us? The counselor knows we're here…." broke in Borodine.

"We've already taken care of that, haven't we, Petersen?"

The secretary nodded.

"He called the counselor, with a message from you, since you hadn't been able to call him directly: "Everything in order with Professor Nudeliev. We're following a lead that takes us back to Alkenia.". Except that your shuttle has already had an accident on the way there…. You know transmissions aren't so safe yet. That's why we're working for you…. No, George, nobody will be looking for you, here or anywhere else."

Nudeliev checked the time.

"We've still got a little time left. But let's be on our way…. I'll bet you can't wait…. It's an enormous honor for you, wouldn't you say?"

They were freed from the chairs but remained with their hands tied behind their backs as they were prodded at gunpoint into another, much larger room which immediately reminded Helias of the time inversion cave on Alkenia, though it was smaller in scale. Here, too, there was a semicircular tunnel constructed of some transparent material, and a waiting shuttle. There was also a large, complicated device in a transparent cabinet against the wall. Helias looked at it, wondering what it was for.

Nudeliev noticed his interest and said "You see, here's where I keep my most precious products. Makes you curious, doesn't it? Yes, we've got time, and I'll tell you what it's for…. But first let me show you what's going to happen to you later, and what your role is. As you may have guessed, it is in this tunnel that the shuttle, with you inside, will go through a self-initiated time inversion, in the sense that the shuttle will do everything itself, as it will again when you reach destination, at the target time. I won't tell you when that is, so as not to ruin the surprise. So everything is programmed and the tunnel is practically inert, no membranes or anything like that. It's used only to delimit the route and absorb the energy, given that we're not in empty space, as you will be for the next inversion. The shuttle is programmed to take you to a given place on Earth at the target time. Once it lands, the harnesses that secure you to the seats and keep you from doing anything foolish will be released, even though there's not much you could do anyway, since there are no manual controls. And so you will be free…. Assuming that everything goes well. As I told you, the whole business is at the experimental stage, and you're the guinea pigs. If you are there and still alive, a gadget on the shuttle will detect your heartbeats and set off a process of self-destruction. But you'll have plenty of time to escape…. Eh, you ought to thank me for this, since whether or not you blow up together with the ship is all the same to me. What matters to me is that, at the moment of the explosion, a particular radiation will be emitted that I will be able to pick up from here even after a lapse of time and determine whether it in fact arrives from the target time. That way I will know that you arrived at the target time alive, and thus that the experiment was successful. Even more importantly, I will know if I am right and you have changed history, because I will be able to check whether this radiation reaches me via the usual four space-time dimensions, or via the seventh dimension. In the latter case, I will know that history has forked and that you belong to a new, freshly created reality, with a history that differs from ours, the respective world lines separated by the seventh dimension, which I will thus be able to discover and study…."

Helias had a question.

"But if this took place in the past, you should already be detecting that radiation. But you aren't, are you?"

"Correct. You're bright, I knew that. No, nothing yet, but we're at the limit of detection. That doesn't mean it isn't there, maybe it's just too weak, but we're still looking."

"Or the experiment failed…."

"That's the other possibility. But I'd swear that this time everything is according to plan…. But come on, why so glum? In any case it will be a unique experience, absolutely extraordinary, don't you think? Something unforgettable…. And of unprecedented scientific reach…. If all goes well…."

Helias looked at Borodine, and Borodine looked at Helias. Each was asking himself, and seemed to be asking the other, if there was any way out. But tied hands and pointed pistols left little play for the imagination. Any headstrong act on their part and they would end up unconscious at best, and thus unable to act or react later.

Better to wait for a more promising opening, if one ever came. And come what may, better not to react impulsively.

"And now we come to that little treasure of mine that piqued your interest. As I was telling you earlier, all of this is frightfully expensive, to say nothing of my friend here, you have no idea how greedy he is and how rich he wants to be. In there is a special antimatter fiber, which cost me an arm and a leg, but it was an excellent investment…. As you know, any electromagnetic signal, of any frequency or kind, travels at a finite speed, the speed of light, and thus takes a certain time to cover a certain distance. For example communications between here and Alkenia, whether they're backwards or forwards in time, travel for more than a year and a half to cover the distance between the two planets, even if, thanks to the little gimmick with retarded radiation in one direction and advanced radiation in the other, we practically don't even notice. When I send a signal, a piece of information, to Alkenia, it gets there a year and a half later. Conversely, from Alkenia to here, the year and a half flows backwards, and so we can communicate 'instantaneously'. But these are all things you know perfectly well. If, however, instead of a signal receiver, I put a reflector on Alkenia that sends my retarded radiation back to me, I'll receive that signal around three years later, and I'll be able to hear, or read, what I wanted to 'tell myself' after a lapse of time. Obviously, I can do the same thing with advanced radiation, it's enough that the devices involved be made of antimatter, thus sending me information back in time, and letting me know what will happen in the future. Well, nobody in their right mind would want to get caught up in a dizzyingly vicious circle like that, with all the risk of getting unpalatable foretastes of your own fate, death included. And in any case, all this is strictly forbidden, and any communication of the kind would be intercepted and punished. But the temptation is great. As you know, for example, lotteries are regularly held on Earth. In other words, you bet on numbers that, if drawn, can win you a fortune. To say nothing of investments in what they call the financial markets: if you know how certain stocks will be trending in the future, wealth is guaranteed. Sometimes it's enough to 'see' a few days ahead. Which means a travel time for advanced radiation that lasts for the same amount. For example, a reflector located in space at a distance of a couple of light-days. But it would be found and destroyed forthwith. And here's where this little contraption comes in: with it, everything takes place in the space of a few meters, in an isolated room, or even by cable as far as my home. The antimatter fiber contained in here is able to reflect any advanced radiation signal ten to fourteen times, which, since it's almost a meter and thirty centimeters long, means a travel time of nearly five days. Enough to do the trick…."

Professor Nudeliev had been punctuating his words with occasional glances at the time. Helias, who had been following the explanation closely, took advantage of a pause.

"Professor, allow me one question. Since you're so interested in bifurcations of reality, why don't you use this little expedient of yours to introduce changes in the history of a few days ago? All you would have to do is send a message to yourself that differs from what you received five days before and, according to your theory, you would have produced a fork in history, without needing to go—or send us, in other words—so far back in time…."

"Dr. Kadler, your question is either naïve or self-serving. I have no intention of gambling with my existence, at least until I'm sure of what I'm doing and have the whole problem under control…. But now let me give you a demonstration of how my 'little expedient', as you call it, works. It's time, Petersen. Send us the information we received five days ago, which netted us such handsome earnings on the earthling markets. And shortly, we will be receiving news of our near future."

Petersen unpocketed his handheld and did as he was told.

"In a few seconds, which are needed for synchronization, the information will depart, and will travel back in time inside there until it reaches the time when we received it five days ago. Actually, we've been indulging in a little theatrics today. It's obviously not necessary to send the message at exactly the right moment. Usually we store it in some moment before or after the transmission, which takes place automatically at the exact time."

Helias found himself wondering whether that information originated in that future moment, or was the information that was already known and received in the past. But he knew that none of this was relevant now, and put on a smugly challenging expression.

"Professor, would you be able to tell me why you don't inform your past self now about what happened today, so that he can take precautions against our presence here and the risks that it entailed and, if I may say so, might still entail?"

But Nudeliev took the challenge in his stride.

"That's obvious. Because it didn't happen. Doing so would introduce a fork, and as I told you, we do not want to do that now…. And anyway, I'd say there's no need, seeing how things are going…."

"No, Professor, you haven't really answered my question. Don't you notice something that doesn't ring quite true?" asked Helias again, eyeing the shuttle.

But Nudeliev had gone back to checking the time.

"Now the information from the future should arrive. A few seconds more…."

Nudeliev was starting to get edgy. Helias's words, after all, had got under his skin. As if a grain of sand had unexpectedly fallen into the delicate mechanism.

Petersen was busy with his handheld, shaking his head. And watching the seconds tick away as he anxiously checked for mail. The time had come and passed.

"Nothing. Nothing's coming." he said, looking at Nudeliev.

Panic flitted across the scientist's face. And Helias was looking very, very pleased with himself.

Then Professor Nudeliev seemed to rally.

"It's probably just a little mixup. A banal delay. Some minor setback. Maybe a glitch with the equipment…. In a bit, maybe…."

"No, Valeri. No glitch, no delay. That message will not arrive. Not in a bit, not ever. Because you will not be here to send it."

It was Professor Borodine's voice. They all heard him.

But Professor Borodine had not spoken.

Something had moved inside the shuttle, whence the voice had come.

It was that white head that Helias had caught sight of a few moments before, half hidden by the seats, and which had raised a finger to its lips, enjoining silence.

Now that head, and the body attached to it, were clearly visible inside the shuttle.

Petersen and Nudeliev had swung around, instinctively pointing their weapons in that direction.

Helias had seized the moment, throwing his full weight onto Nudeliev who, in falling, knocked Petersen off balance. Reacting to the perceived attack, Petersen turned and, before falling himself, fired off several wild shots, hitting Nudeliev in a foot. When the figurative dust settled, all three of them were on the ground. Nudeliev was groaning, whether because of his foot or because of Helias's weight pinning him down was far from clear. Petersen was groaning too, nursing the gun hand that a well-placed shot had put out of action.

The shot had been fired by Professor Borodine the Elder, who had come out of the shuttle and was now standing over them. Seen from below, he seemed taller as well as older, and decidedly thinner.

Professor Borodine the Younger had followed the whole scene with astonishment. And now he was looking at himself, wild-eyed and mouth agape.

"Hi, George!" he heard his other self say. "Buck up, it's all over. For now…."

Then, holding the other two at gunpoint, Borodine the Elder freed Helias's hands so that he could rise and untie the professor. After that, Helias picked up Nudeliev's pistol and kicked Petersen's out of reach.

For lack of a chair, Borodine the Younger had sought out a wall to lean against. A position from which he continued to gape half-wittedly at his older, slimmer and fitter self, a dynamic and confident self.

As if he had read his thoughts, the older self turned towards him.

"It's about time you stopped stuffing your face every chance you get, George. And got a little exercise, don't you think?"

And he winked at him as he patted his now-unprominent paunch.

He then turned to the two accomplices, who had got to their feet under Helias's orders.

"And so you wanted to send us off for a little outing in the past, who knows where. And who knows whether we would have got there in one piece…. Right…. You know all the sayings earthlings have for situations like this? Things like 'giving tit for tat', or 'as ye sow, so shall ye reap', or the even pithier 'live by the sword, die by the sword'…. It seems to me you found it highly amusing. Exciting, even. We were supposed to think ourselves honored to participate in such an experiment. Isn't that right? Well, now the honor is all yours, as is the unforgettable experience…. You don't like the idea, Valeri? Come on, don't be such a big baby. What do you think you're up to in that pocket? Trying to send yourself an SOS in the past? Turned on the mic in your handheld in order to send yourself the whole soundtrack of what's going on? Or even turned on some kind of videocamera for the same reason? And so you'd abandon your admirable resolve not to bifurcate your own past? Hmm, then it's true, as they say, that 'fear is a fine spur'…. Though to tell the truth, I'd hardly say you need to be spurred to do something you have no choice in…. It's all useless, Valeri. I've deactivated your time thingamajig from the shuttle's computer. And while I was at it, I made a few little changes to your travel program, putting in a nice new destination. So you'll have a surprise…. And why so glum now? You're thinking I'm taking my revenge too far? But it's not revenge, not at all. You know, even we up there, in the future, are interested in the seventh dimension, though for entirely different reasons. And this is an excellent opportunity for us. Too bad if it comes at the cost of a few little sacrifices…, of insignificant human beings…, isn't that right, Valeri? But never fear, there won't be an explosion, I've seen to that too. What will be emitted is a much more innocuous signal, but one that's no less effective. And indeed, we've already detected it, though admittedly with a few unexpected anomalies…."

The professor had broken off, his attention drawn by the sound of the shutters rolling up behind him.

In that split second of distraction, Petersen saw his last best hope and threw himself against the professor. Nudeliev, an instant later, went for Helias.

But Helias had remained vigilant and, after immobilizing Petersen with a scatter beam, had stuck his pistol to within an inch of Nudeliev's nose, a nose that, along with its owner, backed off immediately.

The professor, unfazed by the commotion, had turned back to them calmly.

"Useless attempt, gentlemen. Which I obviously knew about already. Don't forget that I've already seen all this…."

And he stole a glance at his 'young' alter ego who, sitting on the floor now, seemed to be asking himself what else could possibly befall him.

"You've got no way out, believe me. I know quite well what will happen in the next, by now very few, minutes…."

He had said it with a note of sadness. Almost as if he had wanted to say 'we've got no way out…'. As if the unknown was about to unfold before his eyes.

Helias watched him in amazement, as a shuttle moved backwards from the open shutters at the far end of the room.

A little voice was heard from the corner where Borodine the Younger had slumped to the floor.

"You're going away?"

"We're going away now, as you can see…." said Borodine the Elder, gesturing toward the shuttle that had just come in. "I didn't tell you before, but I have to leave too…. I'm the only one who can activate that signal and, from there, understand the anomaly…. But there's no time for explanations now. Come on, let's go!"

Then, turning to the huddled figure, Borodine added "Be seeing you.".

As they shepherded the two prisoners at gunpoint, the professor handed Helias a card.

"Here are all the changes I made to the travel program, including our destination, as well as the password for getting out of here. You'll find a copy of the original program and all the operating software in the computer back there in the lab."

The professor's voice had grown solemn and Helias, not knowing what else to say, replied with a stupid "Thanks.".

"Oh, Helias says to say hello…." added Borodine.

Which prompted Helias's second idiotic answer.

"I know…." he said, more or less to himself, thinking of the encounter of the evening before.

Suddenly, Nudeliev fetched to a halt ahead of them.

"B…but there's nobody in there…." he stammered, looking at the incoming shuttle.

"Who's to say?" commented Borodine. "Maybe we're just transparent. As we obviously would be, if we've turned off the interior lights. You ought to know that, Valeri…."

"Why would we have turned them off? There's nobody there, I tell you! The same thing is happening as with the guinea pigs, who disappeared…."

"There's no time to quibble about it." the professor cut him short. And he aimed his pistol at Nudeliev's head, firing a beam that left him docile as a little lamb.

Then it was Petersen's turn to dig in his heels.

"I'm not coming. You can kill me if you like, but I'm not moving from here…."

"Ah, for what little good you are to me, you might as well stay here…."

Borodine shot again, and Petersen crashed senseless to the floor.

The other shuttle was slowly approaching the other mouth of the tunnel.

The professor shoved Nudeliev into the waiting shuttle and strapped him to a seat.

Helias had followed them as far as the door.

"Professor, why are you doing this?"

"I've already done it. The signal has already been received."

"That's not an answer."

"That signal is important…."

"So important as to risk your life? You've already saved ours, isn't that enough for you?"

"I'm glad you see it like that and that you're trying to keep me here, but as you can see…."

And Borodine glanced over at the shuttle that was now entering the tunnel.

"As far as we know, it could be empty."

"No, it isn't."

"You haven't answered me yet…."

"I have to understand that anomaly…, but there's more to it than that, you're right."

"Professor, you haven't already seen these things, have you?"

Borodine gave Helias a long, hard look.

"How do you know?"

"Never mind, just answer…."

"The last time I was here, I was trussed up like a chicken on this shuttle, with you…. This is a bifurcation, Helias. No, I'm not he, I'm not the one caught up in these events. I'm the other one, on the other branch of the fork, sent with you in the past, to that explosion triggered by this madman, with all the damage to the dimensional fabric that ensued. But as you can see, we did all right, thanks to you, though you don't know it yet. Now I'm going back there, to try and sort out that enormous mess, and I'm taking him with me, since he'll be able to give me a hand.

But this time I don't know if I'll ever be coming back.... He made a real mess.... Too much of a mess...."

The elder Borodine had pronounced the last words almost inaudibly as he strapped himself to the seat. Then he had drawn his pistol and fired straight at Helias's forehead.

"Sorry, but you must not remember these confidences of mine."

Helias had a moment of confusion, and then repeated "You haven't answered me yet....".

But the shuttle had already left.

Chapter 17
Helias Had Sat Down on the Step

Helias had sat down on the step that marked the edge of the shuttle's route, head in hands, and with the sensation of having forgotten something important. Then he had raised his head and looked over there, to his right, toward the center of the transparent tunnel. Just in time to see the two shuttles meet and melt into nothingness. As if both had passed beyond the surface of the mirror that divided them, and that now had nothing left to reflect. There was only a fleeting flash of light, then nothing. Nothing remained to testify to the recent tumult.

Then Helias heard a voice behind him, rousing him from his torpor.

"He's gone…. Why did he go away?…"

Helias opened his mouth to answer. But then he realized that he had no answer. It was as if he knew, but wasn't able to remember. He had the sensation that the professor, before leaving, had told him something. Something that, somehow, who knows how, had been erased.

"How… how old was I, do you think?"

Helias turned, and found the professor seated on the step next to him.

"Who can say? I don't even know how old you are now…. Let's say a dozen years older."

"What would you say, was Nudeliev right when he said there was nobody on the other shuttle?"

The professor had the air of a frightened child who, faced with big questions and lacking the courage to find his own answers, was looking for comfort and confirmation from an adult. But Helias was too tired to accept that role. The only thing he was able to say was "You're the expert…."

Then he handed the card to the professor.

"Part of it is here. You ought to be able to find the rest in the computer back there."

On taking the card, the professor perked up and hurried off towards the lab. Now he had something that would help him find the answers he was looking for.

Helias got to his feet too. And slowly began to secure Petersen's hands, though the man still showed no signs of coming to.

M. Villata, *The Dark Arrow of Time*, Science and Fiction,
https://doi.org/10.1007/978-3-319-67486-5_17

Chapter 18
Everything Had Ended Well

Everything had ended well. Almost everything. At least as far as this present of theirs was concerned. The future, or their possible futures, was something they'd deal with when the time came.

In the meantime, the professor, who had copied everything copiable from Nudeliev's computers, was passing his days picking through the information in those files, analyzing it all, with brief pauses for spartan meals, and holding his remaining appetite in check with dutiful trips to the gym, emerging sweaty but smiling, and thumbing his nose at the scales.

Helias had finally been reunited with his parents, to his great joy. But there was no trace of Kathia. Something, someone, had slipped through the net that had closed around Nudeliev's secret project. And Kathia's disappearance was part of this something.

All they could do was wait for Mattheus to recover completely. Perhaps, among the information he had collected while he was held captive, they would find the end of the tangled skein that, unraveled, would lead them to Kathia.

Helias's parents were under considerable strain. On the one hand, there was the relief of regaining their freedom and the happiness of finding their son again; one the other, all the doubts about their future, since they still had very little idea of what to do with themselves. Whether to return to Earth or go to Alkenia as earthlings, or stay on Thaýma as Thaymites. Whatever they did, they would have felt lost, cut loose and rootless, without a past that could guide them in planning a future. But it was only a question of time, as Helias knew, they would soon find their way, a new way.

Helias, for his part, had no doubts. He knew where to go. To the only possible place. Where he would still be able to feel her presence.

And that evening, at dusk, seated on the rock surrounded by snowy mountains, it was as if he could see her emerge from the lake again, dripping and heedless of the cold, as if a warm sun still shone on those shores.

And for a moment, had the sensation of hearing her voice. But he couldn't make out the words.

M. Villata, *The Dark Arrow of Time*, Science and Fiction,
https://doi.org/10.1007/978-3-319-67486-5_18

Helias lay back on the rock. And saw Nasymil, already sparkling in the sky. Then he closed his eyes.
It was as if a hand had stroked his forehead.
Or maybe it was only the evening breeze, gently caressing his face.

Appendix
The Science Behind the Fiction

The concept of time travel has always fascinated scientists and writers, as well as philosophers and the lay public. There is an enormous literature about it, starting with ancient myths and legends, sacred Hindu and Buddhist texts, down to modern science fiction novels and movies.

From the scientific standpoint, whether time travel is possible is a question that has sparked a great deal of debate. Often the possibility is ruled out a priori, if only because of the logical and physical paradoxes that traveling in the past would entail. Of the various kinds of "time machine" that have been suggested, those that make use of general relativity concepts and space-time warps (at times together with quantum physics), like closed timelike curves and wormholes, are currently popular. Of the many studies in this field, mention should be made of the seminal paper by Morris, Thorne and Yurtsever.[1] Here, our idea of time travel is very different from those that are now in vogue, and relies on elements of classical and quantum relativistic physics and electrodynamics. Like any other time machine, it would be very difficult, if not indeed impossible, to actually construct. But it is a useful, instructive and intriguing thought experiment that enables us to probe the effective possibility of traveling in time, with all its contradictions and paradoxes. Among other things, our theoretical vision also permits "instantaneous" interplanetary trips, despite all the well-known difficulties—especially in terms of time and energy—that such trips could involve, and that science fiction writers have always struggled with, often inventing highly imaginative solutions. But let's take one thing at a time, beginning… from the beginning.

Traveling in space means moving from one spatial position to another. Similarly, time travel hinges on the idea that we can do the same thing from one time period to another. Obviously, there are major differences. For example, moving in space takes a certain amount of time, from which we can deduce the speed of movement (space/time): the higher the speed, the less time will be taken. Time, though, flows on its own, independently of our will and our actions, unstoppable. We cannot stand

[1]Morris, Michael S.; Thorne, Kip S.; Yurtsever, Ulvi (1988). "Wormholes, Time Machines, and the Weak Energy Condition". Physical Review Letters. 61 (13): 1446–1449. Bibcode:1988PhRvL..61.1446M. doi:10.1103/PhysRevLett.61.1446.

M. Villata, *The Dark Arrow of Time*, Science and Fiction,
https://doi.org/10.1007/978-3-319-67486-5

still in time. Indeed, our existence is a continual trip in time, but a trip whose "speed" we cannot change. Or rather, we have nothing to compare it to, nothing we can use to gauge the "speed" at which we are moving through time. All we can say is that we are traveling in time at a rate of one second per second.

Actually, this is not entirely true. We know from special (and general) relativity that time does not flow in the same way in all reference frames. Because of the so-called "time dilation", a clock will be slower if it is located in a reference frame that is moving relative to the one where we are observing. There, everything slows down. And so we can, in fact, compare the time passing there with the time that passes here, and say, for example, that time is running at half a second per second there, and that our twin there is aging more slowly than we are. Obviously, our twin can say the exact same thing about us, since the situation is perfectly symmetrical. This is the famous "twin paradox", where each of the two twins should be younger than the other. The paradox can be resolved through general relativity, which deals not only with inertial reference frames but also with those affected by acceleration or subject to gravitational fields. Thus, if we want to really compare the age of the two twins after a certain period of time, we have to bring them into the same reference frame—the Earth, say, which is a quasi-inertial frame. One twin remained there, while the other left on a spaceship that accelerated at a very high speed, comparable to that of light. Then, with other velocity variations (slowing and reversing), he returned to his twin brother on Earth. The brother has aged thirty years, while only three have gone by for the astronaut—apparently because he underwent accelerations but the other didn't. As general relativity tells us, acceleration and gravity are equivalent. In fact, time dilation can also take place in an intense gravitational field, as exists near a black hole. Like in the movie "Interstellar", where the characters who approach a black hole for a few hours return to the spaceship to find that their companion has been waiting for them for more than twenty years…

And so, an astronaut can return to the starting point after experiencing a different flow of time, and find everything older than would normally be the case. In theory, he can "travel" as he likes in the future of the place he started from, but only thanks to the accelerations, gravitational or not, he undergoes.

But then, there's nothing really very strange in all this. After all, the theory of relativity has given us over a century to get used to the idea that time can pass in different ways under different conditions, and so "trips" forward in time don't frighten us. Trips backward in time, on the other hand, seem much more intriguing, with all their logical and physical paradoxes.

But again, we'll take one step at a time, following the logical sequence used in this novel.

The question is: is there a limit speed, that can't be exceeded, for a trip in space? Many people would answer: yes, certainly, it's the speed of light! Right… But now we'll change the question: what is the minimum time it takes to cover that distance? Some people would answer: the time taken by light! Wrong… It all depends on who measures the time taken and the space—the distance—covered.

Suppose we want to go to a planet 20 light-years away from Earth. At non-relativistic speeds, say, less than 1/20 of the speed of light (which from now on we'll call c, as all physicists do, and which is around 299,800 km/s), it would take us more than 20 times the time it takes light, or in other words over 400 years. As we increase the speed, however, the relativistic effects of length contraction and time dilation begin to make themselves felt. For an astronaut traveling at extremely high speeds, near that of light (but which can't be reached with today's technologies), 0.995 c for example, the distance from Earth to the planet would be reduced by around a factor of 10, and he would thus take around 2 years to get to it. On Earth, we can calculate that the trip ought to take a little over 20 years. "Earth" years. So why do the two calculations give such wildly different results? Because time dilation means that time on the spaceship runs around 10 times more slowly than it does on Earth, and so the astronaut will reach the planet after only 2 of "his" years.

So what happens if we increase speed further, up to a limit tending (as mathematicians say) to c? Well, that's easy: all the distances tend to zero. And so the time taken by any trip also tends to zero. Seen from this impossible spaceship, any trip, even if it crosses half the Universe, goes by in a flash, an instant. And in fact we would "calculate" that the hands of the clock on board, during those billions of years that the trip lasts from our point of view, won't even budge.

One thing is certain: a spaceship like this will never exist. But we know that there is something that can (and indeed must) travel at speed c: an electromagnetic wave in vacuum, or rather, its quantum correspondent, the photon, or any other hypothetical particle with null mass. The main difference between the photon and the impossible spaceship is that the latter, as speed increases, would continue to gain inertial mass (by the same factor as the time dilation), which tends to infinity as speed tends to c. That's why this spaceship is also theoretically impossible. The massless photon, by contrast, *must* travel at speed c, whatever the reference frame. It's compelled to do so. And so it's usually said that there's no such thing as a reference frame where the photon is at rest. True or not, such a "system" certainly wouldn't have much physical meaning: in addition to hosting a massless particle at rest, all distances would be reduced to zero—no space, in other words. We know, moreover, that the photon's "proper time" is null, just like its mass—and so no time. (In any case, why would a photon need time if it doesn't have anywhere to go?) It would be a "system" we can save only by a little sleight of hand with zeroes and infinities. Since speed equals space/time, it would be 0/0 for the photon at rest, or in other words indeterminate and thus compatible with c, as it must be in any reference frame. Seen from a "normal" reference frame, the distance in space covered by the photon and the time it takes would both be 0 multiplied by infinity (the usual conversion factor, called the Lorentz factor, becomes infinite) and both would in turn be compatible with any finite value. Consequently, their ratio can indeed be equal to c, as required.

Be that as it may, in any case, if you travel at speed c, distances shrink to zero and time doesn't exist. But you have to have null mass, otherwise it can't be done.

And so, as the novel asks, "How could Helias Kadler and the other travelers, with the whole spaceship and its far from negligible mass, 'ride the light'?" The question arises because the novel features interplanetary trips at speed *c*, complete with passengers for whom time doesn't pass and who arrive at their destination in a "flash".

The answer is fairly simple. Let's take the particle-antiparticle "annihilation", with an electron and a positron, for example. They both exist up to the moment of the "collision", after which they disappear and in their place high-energy electromagnetic radiation is released, typically two photons, which obviously travel at speed *c*. So all you need do is have the spaceship interact with an equal amount of antimatter, and presto chango: as a product of the annihilation, you'll have radiation directed at speed *c* towards the chosen destination. Upon arrival, you "simply" use what's known as pair creation, where two photons produce a particle-antiparticle pair, to separate matter and antimatter, and abracadabra, you have your spaceship and passengers again, safe and sound at their destination. Naturally, this assumes that you've got a technology that can restore the original organization of the individual particles. Or that in reality true annihilation doesn't take place, but matter and antimatter, however conjoined, maintain their identity during the null time that the trip lasts.

All of this, if it could be done, would make it possible to "transmit" passengers over arbitrarily large distances in literally no time at all. But for observers on Earth the trip lasted, in our case, twenty years. If Helias Kadler were to decide to come right back to Earth, he'd find everyone had aged forty years, just as in the twin paradox. Isn't there any way to get around this problem? Indeed, there is, "There's the trick", as Mattheus Bodieur laconically remarks in the first chapter of the story. "What trick?" asks Helias Kadler, and we join him in asking.

Let's consider a different, alternative, interpretation of antimatter: antimatter is nothing other than ordinary matter going backward in time. This was being said as early as the 1940s, by Stueckelberg[2] during the war and Feynman[3,4] a few years after, but it attracted relatively little notice, perhaps because physicists then were more concerned with nuclear questions, the atomic bomb first and then the H-bomb. The concept has recently been taken up again in order to explain some of the fundamental mysteries arising from the observation of the cosmos, and "dark energy"[5,6,7] in particular. The scientific arguments presented from this point on,

[2]Stueckelberg, E. C. G. (1942). "La mécanique du point matériel en théorie del relativité et en théorie des quanta". Helvetica Physica Acta. 15: 23–37.

[3]Feynman, R. P. (1948). "A Relativistic Cut-Off for Classical Electrodynamics". Physical Review. 74 (8): 939–946. Bibcode:1948PhRv...74..939F. doi:10.1103/PhysRev.74.939.

[4]Feynman, R. P. (1949). "The Theory of Positrons". Physical Review. 76 (6): 749–759. Bibcode:1949PhRv...76..749F. doi:10.1103/PhysRev.76.749.

[5]Villata, M. (2011). "CPT symmetry and antimatter gravity in general relativity". EPL. 94 (2): 20001. doi:10.1209/0295-5075/94/20001.

[6]Villata, M. (2013). "On the nature of dark energy: the lattice Universe". Astrophysics and Space Science. 345 (1): 1–9. doi:10.1007/s10509-013-1388-3.

[7]Villata, Massimo (2015). "The matter-antimatter interpretation of Kerr spacetime". Annalen der Physik. 527 (7–8): 507–512. doi:10.1002/andp.201500154.

however, are largely the fruit of the author's own heretical mind, or otherwise depart from the views held by mainstream physics. But to paraphrase T.H. Huxley's famous observation, "Every new theory begins as heresy, then becomes doctrine, and ends as superstition."

The physical process that describes electron-positron annihilation, mentioned earlier, is part of quantum electrodynamics (QED), which is a very complicated business. One of Feynman's great merits was that of representing the complex (and largely incomprehensible) formulas for these quantum interactions in simple and instructive pictorial form, called, unsurprisingly, Feynman diagrams.[8] In the case of electron-positron annihilation, the simplest Feynman diagram is as shown in Fig. 1.

In the initial state (bottom), we have one electron (e^-) and one positron (e^+) which approach, and then interact and disappear, with two photons (γ) appearing in their place. This is the common interpretation. But following the path indicated by the red (gray in the printed version) arrow, the alternative interpretation tells us that there is only one electron, coming from the left, which by emitting two photons inverts its "course" in time, so that it can be observed at earlier times as a positron, while simultaneously, as an electron, it was going forward in time.

To visualize the process of electron-positron pair creation, all we have to do is invert the direction of the time axis (putting the past at the top and the future at the bottom) or turn the diagram upside down, exchanging e^- and e^+ (or inverting the red arrow). In this case, the alternative interpretation tells us that there is an electron coming from the future (and thus observed as an e^+), which by absorbing two photons performs a time inversion and goes back toward the future.

And all this is in agreement with the CPT theorem in relativistic quantum physics, which states that if we apply simultaneous transformations of charge conjugation C (a particle becomes its antiparticle and vice versa), parity P (inversion of the spatial axes) and time reversal T to a physical process, we obtain another physically possible process, or in other words one which obeys the same physical laws. This CPT transformation is what we did to pass from annihilation to pair creation.

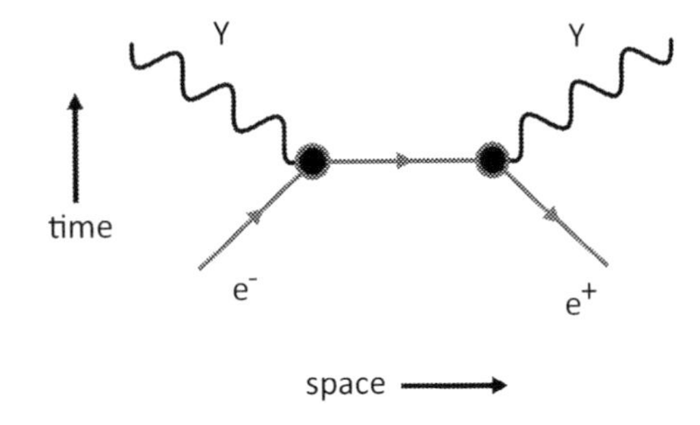

Fig. 1 Feynman diagram of electron-positron annihilation (https://en.wikipedia.org/wiki/Feynman_diagram#/media/File:Feynman_EP_Annihilation.svg)

[8]Kaiser. David (2005). "Physics and Feynman's Diagrams". American Scientist. 93: 156–165.

But perhaps there's a more subtle meaning in the CPT transformation. If the laws of physics must be invariant for CPT (the so-called CPT symmetry of physical laws), and we consider a physical system consisting only of matter and the equations governing its behavior (electrodynamic or gravitational, for example), what does applying CPT transformations to this system in these equations mean? It means describing the behavior of the corresponding system of antimatter as observed from a completely inverted space-time.

The fact that CPT symmetry is the only universally valid symmetry (according to our current physical knowledge), or in other words valid for all types of interaction, including strong and weak nuclear interactions, tells us that these three operations cannot be separated, but must always take place together. So dealing with antimatter also means that we are observing a system that evolves in an upside-down space-time.

Accordingly, the interpretation of antimatter as matter going backward in time (and in space) is practically immediate. We thus discover that there is probably a totally inverted space-time, where ordinary matter can evolve contrariwise, and the existence of antimatter would prove that this inverted space-time is possible.

We know that many of the most familiar physical laws are invariant for P and T (together and singly). This is true of such laws of classical physics as Maxwell's equations for electromagnetism and the laws of gravitation, both in the simplistic Newtonian formulation and in Einstein's relativistic formulation. Maxwell's equations, as well as considering the usual electromagnetic waves traveling in a vacuum at speed c, or in other words, the so-called retarded radiation that propagates forward in time, also contemplate the possible existence of advanced radiation propagating backward in time, though this is usually considered merely a mathematical curiosity with no physical meaning. In our interpretation, however, it is the radiation emitted by matter going back in time.

For gravity, neither the classical nor the relativistic formulation distinguishes between the two time directions. Running backwards a film of a perfectly elastic ball bouncing on the ground or of a satellite orbiting its planet would show nothing unexpected. But it would be different if the ball were not perfectly elastic (balls never are in reality), such that each successive bounce would lose a bit of kinetic energy to heat or deformation, and the backward film wouldn't work. And this is what happens whenever we're dealing with thermodynamics and entropy: a film of a broken glass on the floor that puts itself back together and jumps up onto the table would give rise to considerable suspicion, to say the least. Not that it's physically impossible. Just highly unlikely.

This is why people don't usually believe that another arrow of time exists. But maybe an antimatter glass, in our eyes, would behave in exactly this way.

In conclusion, then, going backward in time is relatively easy (in theory…). All we need to do is what the electron in Fig. 1 does: we transfer an amount of energy equal to twice our mc^2, and there we are, traveling merrily backward in time, and maybe waving to ourselves, our other selves who still have to be inverted. Obviously, in the moment when we "belong" to the other arrow of time, the energy we transferred for inversion is interpreted as energy gained, because now the Feynman diagram is overturned. And the people in the lab where all this happened saw a matter-antimatter annihilation, while for those of us who have been inverted it's a pair creation.

Clearly, there's a problem. There always is when you go backward in time.

Let's suppose we've been inverted, together with the spaceship in which we went through the inversion process. Now, for those who remained faithful to the usual arrow of time, we're made of antimatter (actually, from their point of view we disappeared with the "annihilation", and so we should say "we were made of antimatter"). We "travel" back a bit in time, enjoying ourselves as we watch what's already happened as it unfolds in the opposite sequence, with people walking backwards, getting out of the shower nice and dry with no need of a towel, regurgitating their meals back on the plate, appetizingly untouched. Then we go to one of those instantaneous interplanetary transmission arrival stations we were talking about earlier. Let's say we're on the planet Alkenia and people arrive, or in other words are transmitted, from Earth. And then comes the "trick" hinted at by Mattheus Bodieur! We take the place of part of that antimatter that separates from the matter, which is to say the spaceship and passengers arriving from Earth, after the instantaneous transmission from there to here. For us, that transmission still has to take place, and we "live through it" backwards. And so we end up on Earth at the moment of their departure. So we can not only move from one planet to another in subjectively zero time, but we can also return to the moment of our departure, or at least the same period, without having to go through the dozens of years that light needs to make the round trip. A pretty good deal!

And caroming back and forth, we could even return to Earth in the past, but no further back than the time when the first transmitting station (which for us is the receiving station) was set up. But in the novel, these little jaunts in the past are strictly forbidden by Thaymite law.

But scofflaws, as always, abound. And not only. The book's plot, in fact, revolves around scheming and intrigues, with the theft of secret plans and software that make it possible to travel freely back in time, to any period.

Once transmitted back in time, we find ourselves in a world that flows in the opposite direction, like before the transmission, a world where we're made of antimatter. So we have to invert again if we want to participate in that world's events, and not just watch them like a film being rewound. That calls for another lab for time inversion, or in other words for annihilation-creation; but now it's all more complicated, because we come from that lab's future, unlike when we were inverted before the transmission. It's a far from trivial problem, which in the novel is resolved with special antimatter devices. The lab staff, with no way of knowing what's about to happen, will only detect a massive loss of energy, and an identical pair of spaceships appearing out of nowhere, one "arriving from the future", and the other "going back to the future".

And this is where we run into the problems that we mentioned earlier, the well-known problems that always crop up with trips in the past.

Let's suppose that before being re-inverted, we saw events that happened, say, on Tuesday; maybe events in which we took an active part, as is the case with the characters in the novel. We're inverted twenty-four hours later, on the Monday preceding those events. We already know where we'll be and what we'll be doing the next day… Our fate, it seems, is sealed, and there's no getting away from it… And so goodbye free will!… Or we witnessed events where there's no trace of our presence; or again, we know that on that Tuesday there will be an attack in which many people will be killed—we know because we come from the future. But we can't do anything about it, we can't be there in a place where we weren't seen, nor can we prevent something that's already taken place.

And this is why people don't usually believe that trips in the past are possible. Or at least it's one of the main reasons. Because it would mean abandoning the idea of free will.

Nobody can say for sure whether free will exists or not. We'll never know whether the "choices" we think we make could have been different. At bottom, the physical laws that govern the macroscopic world and thus much of our brain are deterministic. Who can say that the "will" to do this or that is not just a presumption-fueled illusion, and that in fact everything does not play out along a predeterminable chain of causes and effects that we have no real power to change? If I type an "s" on the keyboard now, and then I scratch my itchy eye, how can I ever know whether I could have done differently?

We can invoke quantum indeterminacy, and the fact that in the world of quanta the evolution of a physical process is never certain, but can only be probable at most. And we can think this gives us some room for manoeuvre, but the "how" would be entirely unknown.

Unknown. And in any case, we still know so little about how our own brain works.

Or we can invoke some sort of abstract entity, possibly not physical, like the "conscience", or the "soul", or whatever, and hang the whole burden of free will on it. But we would be hard pressed to say where along the evolutionary ladder from the amoeba to *Homo sapiens* such an entity was injected into the organism. Unless we deny the theory of evolution, and there are those who do just that.

But we've gone well beyond the physical explanations we set out to provide. And all of this, anyway, is entirely outside our own wheelhouse.

Let's go back, to whether it's possible to choose between doing or not doing what we saw ourselves do; or between foiling or not foiling an attack that has already happened. The latter case is easy: clearly we can try, it's just that we don't succeed. It's not as if there was news footage somewhere showing whether or not we were there with our rental car trying to block the road the terrorists are going to take, but getting to the intersection one second too late, or not getting there at all because the traffic cops flagged us down.

But on to the harder case. The one where we not only know exactly what's going to happen down to the last detail, but since we're protagonists, we have to repeat exactly the same actions that we've already seen. Here, it really would seem that there's no room for free will. It would seem that our destiny is that of well-programmed robots who can do absolutely nothing that's not written in their program. Who can't even dig in their heels and say "No, I'm not going there…".

Let's suppose, though, that there's some way we can introduce a change of some kind, large or small. Then we would have two different realities at one and the same time, as it were. With two different futures. Parallel universes? Where only one is the universe we knew and that we come from, while from that moment onwards we'll belong to another? As in the "grandfather paradox", where the grandson goes back in time to when his grandfather was still a boy, and kills him before he can conceive his father (bloodthirsty but effective, more than any other birth control method). And so, after the dire deed is done (or maybe even before, since going back in time), he finds himself in a different universe than the one he came from, and in whose future he'll never be born.

The novel explores all of these different alternatives, where free will may or may not exist, but the view that seems to prevail at the end is that of "bifurcations of reality", where actions in the past give rise to new stories in parallel realities. All of these musings on time travel paradoxes, and others in the novel, spring from the author's speculations, but are often borne out by existing theories,[9] as in the case of the Novikov self-consistency principle.[10]

Or perhaps Professor Borodine was right. When he said that what happens in each instant is, yes, the effect of past causes, but is also the effect of future causes: all you have to do is shake off a single-time outlook on reality. Maybe then there would be no more paradoxes, no more parallel universes. Maybe even the probabilistic nature of the quantum world would gain a meaning: it exists only because we don't take the future causes into account.

Lastly, the novel also deals with other issues, which we might call less scientific, like mind reading and healing powers. "Less scientific" because there is no solid scientific proof that such faculties exist. On the contrary, a number of experiments appear to indicate that they don't, at least for the participants in the experiments. Obviously, this doesn't mean that other people (who maybe weren't interested in the experiments in question) do not have these gifts.

After all, there's also no scientific proof that time travel can exist—far from it—but the scientific literature on the question is extensive and authoritative. So where's the difference? Given that in all three cases there do not appear to be ironclad arguments against them.

The human body, or even any other animal or non-animal organism, is a marvelous machine that can interface with the outside world in ways that are truly incredible. We almost never notice, because we're used to it. A couple of extra faculties like reading people's thoughts or being able to heal wounds would be very

[9]Lewis, David (1976). "The Paradoxes of Time Travel". American Philosophical Quarterly. 13 (2): 145–152.

[10]Friedman, John; Morris, Michael S.; Novikov, Igor D.; Echeverria, Fernando; Klinkhammer, Gunnar; Thorne, Kip S.; Yurtsever, Ulvi (1990). "Cauchy problem in spacetimes with closed timelike curves". Physical Review D. 42 (6): 1915–1930. Bibcode:1990PhRvD..42.1915F. doi:10.1103/PhysRevD.42.1915.

small beer indeed, compared with what we've already got: eyes, ears, the immune system, cell regeneration and all the rest. It might be objected that eyes and ears exist because there are electromagnetic waves and sound waves that, if correctly received, transmitted and interpreted by the brain, provide information of vital importance to the organism. In other words, they exist for reasons of selective pressure.

But the brain emits signals too, in the form of brain waves and so forth. And there are a number of methods for "reading" these signals, the best known being electroencephalography and functional magnetic resonance imaging. Computers can analyze and interpret these signals: it probably won't be long before we have machines that are able to detect whether we're thinking of an apple or a banana. Maybe the difference between these signals and those involving light and sound lies in the fact that thought and brain activity in general have not been around for long, and perceiving them is of less importance to survival (unless somebody is sneaking up on us from behind with lethal intent), and so there has not (as yet) been any real selective pressure in this direction.

The above-mentioned brain imaging through functional magnetic resonance reveals what regions of the brain are activated at a given time, thanks to the increased flow of oxygen-rich blood, which is closely related to neural activity. From the information on the activated brain areas, dedicated software can reveal something of the actual thoughts of the brain owner. This is a very indirect route to mind reading, but it is what we have actually got so far. It would be very different if some people were endowed with the appropriate sensory apparatus, something like the third eye of dharmic spiritual tradition, possibly located in some layer of the neocortex or below. This is, of course, purely speculative, but is a very good topic for fiction.

In addition, the human body emits plentiful infrared radiation. Who knows, in some individuals certain components of this radiation might have beneficial properties. Or there could be some other kind of "radiation", that we've never even dreamed existed because we don't have the right "detectors". Just like when we didn't suspect that there could be such things as radio waves, X-rays, alpha, beta and gamma radiation, neutrinos or gravitational waves.

On more scientific grounds, there is evidence of mind-body connections where some induced mental state can affect healing. The most famous example is the placebo effect. In this sense, a healer would be a person capable of inducing positive and reassuring moods against the disease, thus helping the natural healing process, just as the consolatory effect of a mother relieves her child's suffering. Moreover, if mind reading is possible in the future, maybe mind influencing to heal is also possible. However, this does not explain why some of us, putting our hand on the belly of our newborn baby who suffers from bowel colic, can (almost) immediately make them stop crying upon the release of a great fart.

But then, as we were saying, why is time travel different? Why can it be classified as a scientific topic, and the other things can't? And who ever saw a trip in time? Faith healers and mind readers, real or presumed, legendary or historical, have always existed; what's more, people often believe in them.

The fact is that time travel is harmless (or at least until it actually happens), while paranormal powers aren't, seeing that there has always been a horde of self-proclaimed magicians ready to prey on the gullible. And so, even if only as a question of public health, serious research in this area does not go down well with anyone. Certainly not with the pharmaceutical companies. Nor with the church, whose "spiritual" enemies, heretics and healers, real or presumed, were burnt at the stake only a few centuries ago.

In the novel, the ability to read minds, and perhaps also certain healing powers, arises from a genetic mutation (this is obviously highly speculative, an invention of the author's imaginative mind). The selective pressure on a prehistoric population that brought this mutation about, and the resulting historical repercussions, will be discussed at length in connection with the time trips in the second novel in this series, entitled "The Other Messiah".

Printed in the United States
By Bookmasters